GARY SHERMAN

THE GEOSPATIAL DESKTOP

OPEN SOURCE GIS AND MAPPING

locate
PRESS

THE GEOSPATIAL DESKTOP: OPEN SOURCE GIS AND MAPPING
by Gary Sherman

Published by Locate Press

Editor Tyler Mitchell

Cover Design Julie Springer

Interior Design Based on Tufte-LaTeX

Publisher Website http://www.locatepress.com

Book Website http://geospatialdesktop.com

Contents

1

Forward

Locate Press is proud to bring this second edition of Gary's book into print. The original title, *Desktop GIS*, published by Pragmatic Press successfully sold out, yet there is increasing pressure to have it back in print.

It is special in another way too. It is the first title being published by Locate Press. As our flagship book for this new venture we are thrilled to have such a well-known and competent author as Gary Sherman. His enthusiasm and inspiration have been a pivotal influence in helping us start our publishing endeavour. Thank you Gary!

We've found that it is particularly sought-after by academics looking for a text book for introducing open source GIS into their courses. Seasoned open source GIS users also have been looking for copies to give to their friends.

The demand for having long term stable access to titles like this won't be going away anytime soon. Open source geospatial technology continues to take root around the world and across sectors. Financial crises tend to encourage open source adoption due to the low price tag. Likewise more organisations need to adopt open standards for interoperability and open source geo applications have proven themselves regularly on that front. So now, more than ever, users are seeking out high quality training material from subject matter experts such as Gary.

The movement toward both cloud computing and mobile application platforms also puts increased stress on those products that cannot operate in a free and open source operating system. Fortunately, most open source geo applications are available across all major desktop and server operating systems. So when you need to move from one operating system to the other, your experience will still be similar to what you will learn in this book.

You are likely to find that this book will fill gaps in ways you didn't expect, from practical everyday tips for selecting software to in-depth examples for completing obscure tasks. I believe this is a special book, without comparison, as few other authors have yet taken on the challenge of covering this broad landscape with significant depth.

Enjoy the book,

Tyler Mitchell
Publisher

2

Preface

Open source GIS is a rich and rapidly expanding field of endeavor. Take a look at the FreeGIS Project website,[1] and you'll see an impressive list of more than 300 applications. With such a wide array of software available, it's impossible for any one book to cover everything. In *The Geospatial Desktop*, the goal is to introduce you to some of the major open source GIS applications that are in active development today. It's a tough proposition to cover each of these to the extent they deserve. Instead, the approach is to introduce you to tools that will get you started with open source GIS and enable you to reach out and expand on your own.

[1] http://freegis.org

You might think this book is a beginner's book. Although it's true that it starts out that way, we move quickly into areas that intermediate and advanced users can profit from. Starting from a simple problem and moving through the concepts of using open source, we'll advance to examples of real GIS analysis.

2.1 *How to Use This Book*

If you are new to the concept of GIS, begin at the beginning. For those of you familiar with GIS but new to open source, the introduction is worth reading, but you should definitely take a look at

Chapter 4, Getting Started, on page 25 for an overview of things to consider.

If you want an overview of what's available in open source GIS, before you proceed take a look at Appendix A: Survey of Desktop GIS Software, on page 299.

Following the introductory chapters, we delve into working with data, digitizing and creating new data, and then doing analysis using open source GIS applications such as GRASS, QGIS, and uDig. In later chapters, you'll find information on scripting and writing your own applications.

Since this book is not a tutorial, we won't go into all the nuances of each application mentioned. We will show you how to accomplish common tasks using the software, and in those cases you'll find a fair bit of guidance.

The appendixes contain information on installing and using some of the applications mentioned in the book. If you need further assistance getting started, refer to websites for the respective projects where you'll find a wealth of information.

[2] http://geospatialdesktop.com The illustrations in the book are presented in black and white. Full color images for each figure in the book are available in both HTML and PDF format on the book's website.[2]

2.2 Versions

The dynamic nature of the open source GIS community was readily apparent during the writing of this book with several projects releasing major versions. Fortunately, the differences between the versions don't significantly impact our illustrations and examples. Where there is a difference, it is noted in the text. For software used in the examples, the following versions were used:

GRASS - Version 6.4.x

Quantum GIS - Version 1.7.x

uDig - Version 1.2.1

For GDAL, GMT, PROJ.4, and PostGIS, you can use the latest versions to work through the examples in the book.

2.3   *Acknowledgments*

I want to express my thanks to those who have contributed to the development of this book. The following reviewed and provided input on the first edition: : Markus Neteler, Matthew Perry, Barry Rowlingson, Tyler Mitchell, Frank Warmerdam, Aaron Racicot, Jason Jorgenson, Brent Wood, Dylan Beaudette, Roger Pearson, Martin Dobias, Patti Giuseppe, and Landon Blake.

In particular, thanks go to Tyler Mitchell and Locate Press for getting this work back into print.

I want to thank my family for their support, encouragement, and patience during the entire process.

This book is dedicated to the memory of my father. While from another era, he instilled in me the curiosity of how things work and what to do when they don't. He taught me much, and for that I am forever grateful.

Gary Sherman
February 2012
gsherman@geoapt.com

3

Introduction

Interest in mapping is on the rise, as evidenced by services such as Google Earth, Virtual Earth, MapQuest, and any number of other web mapping mashups. These are all exciting developments, yet there is another realm you should consider - the world of desktop mapping with open source GIS (*OSGIS*). You may be thinking, "Why do I need OSGIS? I have all the web mapping sites and tools I could ever need."

To answer that question, let's consider our friend Harrison. He's coming from the same place as many of us, having played around with web mapping tools, and is now ready to start adding his own data. Harrison quickly discovers he can't add the GPS tracks from his last hike to any of the "conventional" web maps. All he can do is view the data they provide. Next, he fires up Google Earth[1] to see whether that will do the trick. He soon finds that with a little digging he is able to get the tracks off his GPS and import them into Google Earth. With a little bit of work, Harrison is now able to display his GPS tracks.

[1] Although there is a free version of Google Earth, it is not open source.

Fresh from his victory in Google Earth Harrison now embarks on his next project, which is the real reason he is interested in mapping. It turns out that Harrison is an avid bird watcher. Not only did he record his trek, but he also logged waypoints at each bird sighting.

With each waypoint he made a few notes about the species of bird, the number of birds observed, and the weather conditions. Harrison has just moved from simply mapping where he walked, to wanting a way to display his bird sightings and analyze his observations. In doing so, he has hit upon the basis of a Geographic Information System (GIS) - linking geographic locations to information.

Harrison ponders his next move - how to get all that good bird information from his trail weary notebook sheets into a form where he can not only visualize it but even ask some questions (in other words, do analysis). Harrison wants to be able to do the following:

- View the locations where he observed birds
- View only the locations where he saw the yellow-bellied Wonky Finch
- Scale his locations (dots) based on the number of birds seen at each location (more birds = bigger dot)
- See whether there is any relationship to the weather and the number or types of birds he observed

Harrison needs not only a good visualization tool but something he can do analysis with. Harrison needs some GIS tools, and of course we recommend open source desktop GIS as the solution to his mapping needs.

3.1   What Is Desktop Mapping?

Harrison has introduced us to a problem that we can solve with desktop GIS software. So, what exactly is *desktop mapping*? Well, it isn't about drawing a map to find your pencils, pens, stapler, and coffee cup. Desktop mapping is all about using software installed on your computer to visualize and analyze data. Not only can it be used to meet Harrison's bird-mapping needs, it can also create hardcopy maps, create data out of thin air (well almost), and examine the relationships between features.

Although it's true you can do all this with proprietary software, we'll take a journey through the open source GIS landscape to see

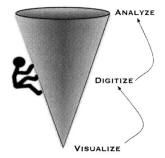

Figure 3.1: GIS functions

what we can find. To get started, let's take a look at the kinds of things we can do with open source desktop mapping tools. I've already told you that Harrison's bird project can be handled quite nicely. Everybody likes to "get on board" quickly rather than learning a bunch of theory and commands. We'll try to do the same here as you start your journey into *OSGIS*, whether you are a beginner or a battle scarred GIS geek. To give you an idea of what is possible, there is a rather simplistic interpretation of things we can do with GIS (open source or not) in Figure 3.1, on the facing page, in order of increasing complexity. We'll take a closer look at each of these functions to help you get an idea of what's involved with each. In turn, you can decide how far up you want to climb. You'll notice that our GIS progression is like scaling the outside of an inverted cone. Imagine yourself as a rock climber doing a free climb up the outside of that cone. The higher you go, the more of a workout you're going to get. Learning *OSGIS* is a bit like climbing that cone. Fortunately, you decide how far to go based on what you want to do. Getting on board is pretty easy. Let's visualize.

Visualize

The dictionary (well, one of them anyway) defines visualize as "make (something) visible to the eye." That definition fits pretty well with what we want to do. We want to *see* our data. This is the entry-level activity in GIS. We get some data, whether from our GPS or by downloading it from the Internet, and we look at it. Remember, that's the first thing Harrison was interested in - looking at his data. That sounds good, but you'll quickly find that looking at a bunch of black lines on a white background isn't very exciting. We need a context for our data. Let's return to Harrison for a moment.

Harrison has caught up with us and is staring at a nice collection of seemingly random dots on a snow white background in his desktop GIS viewer.[2] Although immensely proud of his accomplishment, it really isn't much to look at and certainly not very enlightening. Harrison wishes he could display his data over the same topographic (topo) map he took with him on his hike. Using his favorite search

[2] At the moment, we're talking in generalities; we'll get to some specific *OSGIS* applications shortly.

[3] http://libremap.org

[4] A DRG is a scanned USGS topographic map, typically available in TIFF format.

[5] The astute observer is asking, what about the projection difference between the GPS data and the DRG? We'll pretend that doesn't exist for the moment.

engine, he begins the hunt for a topo map. Fortunately for Harrison, he stumbles upon the Libre Map Project[3] that has free topo (DRG[4]) maps for the United States. Harrison quickly finds his part of the world and downloads the appropriate maps. Now he can overlay his GPS data on a background that provides some context.[5] Harrison gets really ambitious now and goes to hunt for some imagery to add to his map. We'll check on him later.

What's the first thing you are going to do when you add your bird locations, fishing holes, or Big Foot sightings to your map? My guess is you'll want to change the color, symbol, size, and any number of other things to control the way it looks. This is another important aspect of visualization - being able to change the way the data looks. We call this *symbolizing* your data. I think it's safe to say that all *OSGIS* viewers provide this ability. Typically you can change the colors, fill patterns, line styles, and marker symbols to get the effect you want.

Figure 3.2: Bird sightings: The bigger the dot, the more birds

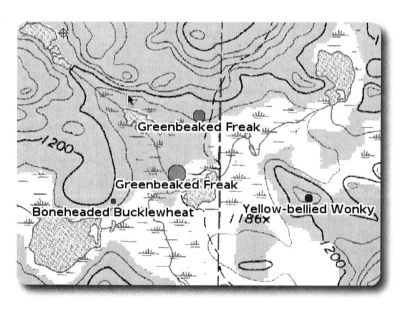

Think back for a moment to Harrison's requirements for his bird-mapping project. He wants to not only view the locations where

he saw the birds, but also change the size of the dot based on how many birds he saw at a location, as in Figure 3.2, on the preceding page. He also may want to display only a single species. Most *OS-GIS* viewers can easily accomplish these tasks and more. Harrison hasn't thought of it yet, but he's going to want to symbolize his bird locations by species as well. By using both colors and sizes, he can convey a lot of information about his birding observations. We'll see examples of how to render our own data using these techniques in *Rendering a Story* on page 44.

GIS Data Types in a Nutshell

You are about to be exposed to a bunch of new terms as we launch into our discussion of what you can do with OSGIS. Among these are GIS data types. Essentially you can divide GIS data into two types: vector and raster.

Think of vector data as things you would draw with a pencil and paper. We could draw points, lines, and polygons. In GIS, the features have a location in the real world, allowing us to examine their relationship to other features.

Taking it one step further, we can attach *attributes*—information about the feature. Our vector data can have one or more attributes. For example, we might create a polygon that represents the outline of a lake. The attributes for the lake might be name, area, perimeter, and mean depth. Attributes are stored in fields in our dataset, whether they be in a file or a database table.

These two characteristics, location and attributes, are what make GIS different from a simple drawing or paint program.

The other type of GIS data is raster data. In a raster, the information is represented by cells (in some cases pixels) where the value of each cell represents a quantity or color. Examples are a photograph where the cells represent a color, and an image where each cell value represents an elevation.

In GIS, we use both types of data, depending on what we are trying to accomplish. In the simplest case we might use a raster image, an aerial photograph in this case, showing our neighborhood. We would then overlay our vector data in the form of streets. It not only makes a nice picture to look at, but with attributes attached to the streets we can also learn the name of each.

Now that we have explored visualization a bit, let's move on to the next step. From visualize to digitize.

Digitize

Let's define what it means to digitize. Breaking out our handy dictionary gives us a definition along the lines of "convert pictures or sound into digital form for processing in a computer." There's nothing mysterious about that definition. See, you may have already done some digitizing when scanning your old photographs or playing with the sound recorder and a microphone on your computer.

When it comes to GIS, digitizing usually means capturing and storing points, lines, or polygons from paper maps. But for the purposes of our general discussion, we'll just equate digitizing with creating data and hope the purists don't catch up with us before we're done.

Harrison has a digitizing project in mind. Looking at the DRG he downloaded to use as a base for his bird visualization project, he finds it shows roads, trails, lakes, contour lines, and other physical features. Unfortunately for Harrison, many of the small lakes on his map are not labeled with their name. To make a better looking display (and ultimately a hard copy map), Harrison would like to label the lakes. He could just use a paint program and label the lakes, but then he would have to modify the original image. Besides, Harrison tends to flip-flop a bit about what he wants so maximum flexibility is important. Harrison then discovers the notion of creating his own *vector* data. If you're not familiar with it, vector data is just points, lines, and polygons that represent real features on the ground. Harrison thinks about creating a point near the middle of each lake and labeling it, but that would look a bit goofy even for him. He then decides to digitize each lake and make a polygon. For each polygon, he'll add an attribute - the name. While Harrison is busy working on his lakes, let's talk briefly about the process of digitizing.

In the simplest sense, digitizing is tracing features with your mouse. In reality, there is a fair bit of skill involved in doing it right. The process goes something like this: you create a new *layer* (think file for now) to store your features in, define some attributes for it (for example, the lake name), and then begin tracing features. As you complete each feature, you enter the attributes. When you are done, you have a layer you can view and label using the attributes you entered. Of course this is a simple explanation, but all digitizing is really an extension of these concepts.

We've kind of lumped things together under the digitizing category. There are other ways to create new data apart from digitizing. Harrison actually illustrated this for us when he imported his GPS data. Other ways to create data include importing from a text file, scanning images, and even accepting coordinates from a web form. We'll get into more of this later. Let's hope by now both you and Harrison have a good idea of just what it means to create GIS data. Once we have all this good data, it's time to analyze.

Analyze

This is where GIS really shines. Being able to use the data we worked so hard on collecting to answer some what-if questions is what makes GIS exciting. This is also what separates GIS software from being just a "viewer."

Using GIS we can answer all kinds of questions. Let's get Harrison to help us out with an example. He has a theory that most of his bird sightings are within 200 feet of a lakeshore. With all his hard work, Harrison can view both his bird sightings and the lakes, but he can't really tell how far apart they are. He could use the fine tools provided by his software to measure the distance from each sighting to the nearest lake or lakes. But this is time consuming and tedious, and the end result can't be visualized. Fortunately, Harrison can use a common GIS operation known as *buffering*.

Harrison proceeds to create a 200 foot buffer around his lakes (see Figure 3.3). This is pretty much a one stop operation. You indicate what layer you want to buffer (lakes) and enter the distance. The software then calculates the buffer around each lake and creates a new layer containing the result. Harrison now proceeds to set up his display. He adds the new buffer layer to the map, then the lakes, and finally the bird sightings. Any bird sighting falling on the buffer layer is within 200 feet of a lake (or lakes). Harrison can quickly visualize his results and see whether his theory is right. OK, so it turns out he was wrong. It looks like the bird sightings don't necessarily fall within 200 feet of a lake. Harrison decides he can still be right and goes off to create a 500 foot buffer.

Figure 3.3: A 200-foot buffer around the lakes

This is a simple example of the type of analysis you can do with open source GIS applications available today. You may be thinking that Harrison's analysis is a bit contrived and really not all that significant, and you are probably right. Let's list a few more situations where a buffer analysis might provide meaningful insight:

- Restrict development to a distance at least 500 meters from an active eagle nest.
- Determine where to allow a drinking establishment so that it's at least a quarter of a mile from any school.
- Develop emergency action plans by identifying all public facilities within a given distance of a hazardous storage site.
- Establish a setback around a creek or stream.

And the list goes on. As you can see, the simple operation of creating a buffer can answer a lot of questions. It's a valuable tool and just one of many that we'll take a look at as we get deeper into specific applications. Of course, there are a lot of other types of analysis we can do with desktop GIS. We'll explore some of these later.

We've now taken a look at three aspects of GIS: visualization, digitizing, and analysis. With that under our belt, we are ready to get into some more specifics. Oh, and about Harrison - he finally proved his point by creating a 5,000 foot buffer around all the lakes. As usual, the tools alone can't provide a meaningful analysis.

Before we move on too far, let's take a quick look at the server side of things.

3.2 *Desktop vs. Server Mapping*

When you think of a server, you probably think of a big machine locked away in an air-conditioned room somewhere. Well that could be true, but in this case I'm referring to software not hardware.

The server side of open source GIS provides important capabilities for us on the desktop. For example, we might have a *spatial database* that stores our data. Or we might have a spatial server that can pump out data using a number of web standards. We can use all these data sources from the desktop.

We might have resources on a bunch of servers, all accessible from our desktop GIS applications and serving up all the data we need. We're still doing desktop mapping, using the tools installed on our local machine. Let's contrast that for a moment with server-side mapping.

You can see an example of server-side mapping by pointing your web browser at one of the many web mapping applications on the Internet. These range from sites providing maps and driving directions to those serving up massive quantities of data. I'm sure everyone has seen examples of the type of applications I'm talking about, but for Harrison's benefit we'll mention a URL. Take a look at the Geo.Data.gov website[6] for some sample web mapping applications.

When using server-side mapping we don't install anything on our local machine, and all we need is a web browser. The good thing about this scenario is that someone else has done all the work in assembling the data and preparing it for display and use. Why would we want to go the desktop route instead of letting someone else do all the work for us? Our friend Harrison discovered some of the reasons in his bird mapping project. He wanted to view *his* data, not the data provided by some server somewhere. He wanted to create new data by digitizing the lakes.[7] Harrison also wanted to analyze his data by buffering and storing the results. A lot of these operations can be done with server-side mapping, but the data ends up living on the server. If you're lucky, there may be a way to export it and make it yours.

[7] Sure, you can create features on some web mapping sites. But where do they reside when it's all said and done? On the server.

Am I down on server-side mapping? No, it's an excellent way to *visualize* data and provide it to the masses. In fact, there are projects underway to further enable the server side and extend the capabilities to analysis as well.[8] Using a mix of desktop and server GIS software is a good mode of operations, especially if you are like Harrison and want to be both a data creator and a consumer.

[8] An example is PyWPS (http://pywps.wald.intevation.org), which allows web access to GRASS GIS.

3.3   *Assembling a Toolkit*

With the preliminaries out of the way, my goal now is to help you assemble a loosely coupled toolkit of *OSGIS* applications. There are good reasons to assemble a toolkit rather than using a single mapping application. Just as you wouldn't use a screwdriver to build an entire house, we'll get better results if we use the right tool for the job at hand. When it comes to *OSGIS*, I'm a strong proponent of *IIWUI* - "If It Works Use It."

Not everyone will need or want the same tools in their toolkit. One of the things we hope to accomplish on our journey together is to identify which tools you need and then learn how to assemble them into a system that works for you. Ideally, you should come through the experience with some nicely integrated applications and utilities to serve all your mapping needs on the desktop. As you assemble your toolkit, you'll find that many applications are of the "Swiss Army knife" variety, providing a wide range of capabilities.

3.4   Other Mapping Options

What are your other options? Well, we already mentioned them - web-based applications. Unless you are developing your own web mapping application, you're pretty much at the mercy of the web developer. You must use their interface and work with the layers they provide. For some folks this is a perfectly good solution, and it's definitely something to consider when you are ready to share your hard work with the rest of the world.

For those of you who need to work with local or distributed datasets to create, edit, and display data, this isn't going to work. You will need tools to create and maintain your data.

A solution that falls in the middle is Google Earth, available on Mac OS X, Linux, and Windows. With Google Earth you can add and display your own vector data once you've converted it to the proper format. I find that using my desktop GIS toolkit to create and prep data for Google Earth meets my *IIWUI* test.

3.5   What's Ahead?

To give you an idea of where we're headed, next we're going to dive into *OSGIS* and look at the whole notion of using open source for your mapping needs. From that point on, we'll look deeper into concepts, data, and use of the tools at our disposal. Our goal is to get you up to speed on working with *OSGIS* desktop applications, and there is a lot of ground to cover. Unfortunately, we can't give you an in-depth tutorial for all the applications we'll use. In the appendixes you'll find some additional information for some of them, and we'll point you to additional resources as we go along.

4

Getting Started

Before you start madly downloading software to assemble your GIS toolkit, let's think a bit about your requirements, including what type of mapping you are interested in doing. You may not know the answer to that question. Most likely if you are starting out, you'll follow the same path as Harrison—moving from visualization to creating your own data to doing analysis. Ultimately, your needs, goals, and requirements will guide you in assembling your toolkit. For example, there is no point in assembling an industrial-strength system to simply view GPS tracks on a map.

As you explore your needs, remember to keep open the possibility for expansion. As you begin your journey into OSGIS, you may end up at a destination you never considered. The good thing is, you can always "upgrade" your toolkit.

4.1   The Three User Classes

If you are already a GIS user, you likely have a good idea of your needs and requirements, but it's always good to reevaluate. Let's consider three classes of GIS users to help you get started. To help us get acquainted, we'll use the names Clive, Irving, and Alyssa.

The Casual User

Clive is a casual user, and what he likes to do is visualize mapping data. His toolkit contains one or more GIS viewer applications and maybe a custom *data store*—a place where data resides—such as a spatial database. In the simplest case, Clive stores his data in files (shapefiles, Tagged Image File Format [TIFF], and so on). He doesn't need big fancy GIS algorithms to make him happy. Clive may on occasion need to create data by importing GPS tracks or maybe even digitizing some lakes or trails.

Since he doesn't create a lot of data, Clive gets it by downloading from the Internet and sometimes from his GPS, just like Harrison did in the first chapter. The other things Clive uses his GIS software for are printing simple maps and doing some visual analysis by plopping layers on top of each other.

The Intermediate User

Irving is an intermediate user, and he likes to not only visualize but to create data—sometimes lots of it. Irving typically creates data by digitizing and/or converting it from other sources. Sometimes Irving needs to produce cartographic output (a paper map with lots of decorations) to share with his friends and cohorts.

Irving works with a wider range of data formats than Clive. He likes to digitize data from raster maps (just like Harrison), convert data to suit his needs, create subsets of his data to better visualize where things are, and use symbols to help visualize some of the relationships between features.

The Advanced User

Alyssa is an advanced user, and she has mastered the activities and tools used by Clive and Irving. But she has greater needs—Alyssa lives to analyze. Beyond viewing, data creation, and map production, she uses GIS to answer questions based on spatial relationships. She does cell-based analysis and perhaps even routing and geocoding. She also may need to write programs and scripts to accomplish her tasks.

Some of the tasks that Alyssa performs include doing line-of-sight analysis ("Can you see me from here?"), change analysis, buffer analysis, and grid algebra. She needs a high-powered toolkit.

4.2   *Which Are You?*

What do Casual Clive, Intermediate Irving, and Advanced Alyssa all have in common? They all started at the same place and they each use some of the same tools. You'll also notice that the classes of users bear a strong resemblance to the functions in our GIS Cone in Figure 3.1, on page 14.

Based on our characters, you should be able to determine where you fall in the lineup. Not only should you consider what you are now, but what your needs will be in, say, six months, a year, and beyond. The truth is that each of our users may have the same tools in their toolkit. The difference will be in how they are used and to what extent. As we progress through our exploration of desktop applications and their capabilities, keep your self-assessment in mind. We'll provide reminders along the way to indicate which tools are best suited for each class of user.

Lastly, this artificial classification scheme is not hard and fast. It only provides a starting point for you to think about your requirements and help you build up your own open source GIS toolkit. Although jumping into the deep end of the pool can be an effective, albeit traumatic, way to learn to swim, sometimes it pays to wade in gradually, feeling your way along. The more feature-rich an application is, the more likely it is to have a steeper learning curve. You start with the tools that meet your needs and work your way into the more complex as your appetite for GIS increases.

Determining what kind of user we are wasn't too bad. Now we move into something a bit more difficult and look at some of the challenges in assembling an open source GIS toolkit. Everything you do (including crossing the street) entails some level of risk.

Whether you use open source or proprietary (closed source) software, you incur some risk. The rest of this chapter looks at the challenges and risks and provides some insight on dealing with potential pitfalls.

4.3   Choosing a Platform

In ancient times (around twenty to thirty years ago), if you wanted to "do" GIS, you had to buy a certain type of hardware running a specific operating system. As time went on, the choices increased. Today you can pretty much find GIS software to run on your favorite system, assuming it's Linux, *nix, OS X, or Windows. You still can't find much in the way of GIS software for your Commodore 64.

The logical assumption might be that we just get the software for our current platform and forge ahead. But consider this: should you choose the software for the platform or the platform for the software? There are a number of factors to consider:

- Your comfort level with various operating systems
- The types of applications you need
- Your budget

Typically you will choose the software for your current platform and be on your way. For those of you who are comfortable in two or more operating systems (say Linux, Mac OS X, or Windows), your options are more varied. I would rank my choices pretty much in that order. If you have a choice, Linux or OS X may be a better fit for you. If not, we plan to show you Windows users plenty of options in the coming chapters.

Budget figures in somewhat, with the hardware for some platforms costing more than others. Since your software acquisition costs are going to be low or nonexistent, you can afford to spend a little more on hardware.

4.4   Selecting the Right Toolkit

We mentioned this earlier, but it's worth repeating. Pick the applications for your toolkit based on what you want to do. There is no point in installing every GIS application out there to view Grandma's house and the local latte stand. Those are valid uses, but why make it hard on yourself? On the other hand, you should think ahead a bit and keep your options open. That way we won't end up installing a simple viewer and expect to do volume or fill analysis. To give you a head start, we'll be looking at the applications shown in Figure 4.1, which also shows the appropriate user classes. This is really a generalization, but it does give you an idea of the level of experience appropriate for each application. In reality, many of the applications can be used across the spectrum of user classes.

To help you learn more about the software choices available, you

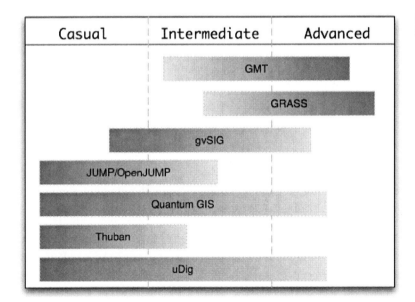

Figure 4.1: Some OSGIS applications in relation to class of user

can refer to the survey of open source desktop GIS applications and some of the capabilities of each in *Appendix A: Survey of Desktop GIS Software*, on page 299.

4.5   Acquiring and Installing Software

Getting an OSGIS package can be a bit different than buying something off the shelf. Although it's true you can purchase Linux distributions off the shelf that include OSGIS software, typically you are going to be downloading a binary package for your platform. If you're in luck, that's the situation you'll encounter as you begin to assemble your toolkit. This route allows you start using the software without worrying about all those nasty things like dependencies, compilers, and libraries. The worst case is you may have to download the source code and compile it yourself, although this is rare in most instances.

A binary package is simply and compiled and ready to run set of one or more applications. You download and install and you're ready to run.

All the OSGIS software we'll see in this book can be obtained as a binary package or installer, depending on your operating system. This is the easiest way to get started using an application. Here are some things to be aware of when going the binary route:

Packages, Installers, and Disk Images - Depending on your platform of choice, you may be installing RPMs, DEBs, or tgz on Linux; zip or bundled installers on Windows; and disk images (dmg) or OS X installers on Mac. Most open source GIS projects provide these binary images, and of course it's up to you to determine which to install. As we go along, we'll give you hints on the installation process and mention all three platforms.

- Some packages and/or installers are not provided or maintained by the open source project but by third parties.
- Depending on your operating system, the latest version may not be available.
- The availability of packages for your platform may lag behind the general release of a new version.

Going with a binary package or installer is definitely the way to go when test-driving an application for the first time. This gives you a chance to easily try things without the hassle of gathering dependencies and compiling from source.

In some cases, you have to compile from source because you have no alternative. Here are some reasons why you might want to compile an OSGIS application:

- The binary isn't available for your platform.
- You want the latest and greatest features, but they haven't been released yet.
- You want to *customize* your toolkit.

Compiling a suite of tools from source can be a daunting task for the average user, even for the advanced GIS user. When first starting out, you should consider using binary packages for your platform. This keeps you from becoming frustrated with the process of boot strapping a system from scratch. Once you gain familiarity with the tools and how they interact, you'll be ready to venture into compiling your own system. For now, let's start with the packaged binaries and learn how to use the software rather than get frustrated with the build process out of the gate.

> ### Trying Open Source GIS with a Live DVD
>
> Another option for giving OSGIS a spin is one of the many Live DVDs. These allow you to boot your computer from DVD into a Linux system that is preloaded with applications you can use without having to install anything.
> You can choose from a number of GIS Live DVDs, but you need to make sure your choice contains fairly recent versions of software. For a Live DVD that attempts to provide the latest versions, check out the OSGeo offering at http://live.osgeo.org.

4.6   Integration of Tools

Rarely will you find one OSGIS application that meets all your needs. In fact, if you do, you're in the minority. An OSGIS toolkit composed of several applications will provide a much more powerful and complete system. Now you're thinking, "Oh, great, I have to learn a whole bunch of new programs to do anything with this stuff." In reality, we'll show you how to get started without a huge learning curve. For those of you already up to speed on GIS and tools, we'll provide that deeper view you're looking for to fill out a complete toolkit.

How Do We Integrate?

The plain fact is that integration is largely up to you. Typically you'll end up with a loosely coupled set of tools, sometimes bound together with scripts or other glue. This shouldn't be interpreted

Kludge: A program or system that has been poorly (perhaps sloppily) assembled.

to mean that we are creating a *kludge*, but rather putting our tools neatly in the toolkit and making them play nicely together. This allows us to create a modular set of open source applications that we can swap in and out as dictated by our needs and experience.

Some tools integrate nicely, and the situation is improving all the time. An example is Quantum GIS (QGIS) and GRASS integration. The GRASS plugin allows you to access a large number of GRASS functions through the QGIS interface.

Another form of integration is using programming language bindings so that you can access the application functionality in Ruby, Python, Perl, and Java programs. We'll talk more about this technique in *Using Command-Line Tools*, on page 199.

4.7 *Managing Software Change*

One of the biggest challenges you will face when using OSGIS software is managing change. All systems have an inherent element of risk with regard to change. Computer systems are particularly sensitive to change, meaning if you upgrade one component, you better make sure it doesn't have a negative impact (read: complete meltdown) on the other components. Let's look at an example.

[1] An imaginary open source GIS application

Harrison hears about some really cool features that were just added to SuperDuperMapper.[1] After all, Harrison subscribes to the project's email list and participates on IRC so he can be "in the know." Unfortunately, the new SuperDuperMapper features are in the development version. Undaunted, Harrison proceeds to check out the source code and build the latest, greatest version. And it works great. All the new features are there, and Harrison is one happy mapper—until he goes to run his faithful old workhorse application, MundaneMapper. Turns out that his hacking activities have introduced some library incompatibilities, and now MundaneMapper refuses to start. Harrison has become a victim of *BES*.

Harrison will glumly tell you that if you want to maintain a stable system, the first thing to avoid is Bleeding-Edge Syndrome (*BES*).

This differs from being an "early adopter." Here is how to tell if you have *BES*:

- You always download and install the latest beta.
- You find yourself doing Git, SVN, and CVS checkouts and building from scratch.
- You subscribe to the Git/CVS/SVN commit mailing lists for several projects and rebuild your toolkit each time a new message comes in.
- You often find yourself with an inoperable system.

Having *BES* is not so bad if you are a hobbyist or just experimenting and understand the risks. It's not so good if you are trying to do real work and can't afford to break things on a regular basis.

Guidelines for Managing Change

Managing change really refers to keeping your software current, responding to security issues, and keeping things stable so your toolkit can serve you, not the opposite.

Let's look at the three main reasons to upgrade:

- A new version has been released that provides features you need, want, or absolutely can't live without.
- Vulnerabilities in your software.
- A "higher-level" component (such as your operating system) requires an upgrade that will render your toolkit applications incompatible.

The first two are a matter of choice; the third may not be if your IT department has any say. If you are lucky, you are master of your own destiny and have control over all aspects of your GIS software, including the operating system. If not, you're going to have to coordinate and cooperate.

Here's a list of some suggestions for managing change in your OS-GIS toolkit:

- Proceed with caution. In other words, look before you leap, and

Git, Subversion (SVN), and Concurrent Versions System (CVS) are version control systems used when developing software.

make sure you understand all the ramifications of upgrading.

- Identify changes in the latest version of the application(s) that may require extra work on your part.
- Identify changes that remove key functionality you depend on (it sounds strange, but I've seen it happen before).
- Identify dependencies—other packages that will break or things you need to upgrade as part of the process.
- If at all possible, test your upgrades on a nonproduction machine (virtual machines are great for testing new releases).
- Don't upgrade too quickly after a new release. Monitoring the mailing lists and forums can help identify potential problems that others have already discovered (and oftentimes, solved).

You may be thinking this OSGIS approach is a minefield. In reality, it's no different from managing change with proprietary software. All of the suggestions mentioned here apply equally to both proprietary and open source software, particularly in the GIS realm. Just be smart and never put your data at risk, and you'll be fine.

4.8   Getting Support

Open source software has a unique support system, and OSGIS is no different. Rarely when using a proprietary application can you communicate with the actual developers—with OSGIS you can, often in real time. Most developers are willing to help, assuming you have spent a bit of time working through things yourself and reading the documentation. Some of the support channels you can use are as follows:

- Mailing lists
- Forums
- IRC (a real-time service that allows you to chat with people across the globe)
- Wikis
- Search engines
- Websites

When using mailing lists for support, you need to be sure to search the archives before posting your question. Quite often the answer to your question will be waiting for you to discover it. In addition to the archives typically maintained by each mailing list, a couple of other independent archives are quite useful: Nabble[2] and Gmane.[3] If the archives don't provide the answer, compose an email to the list, and make sure you include enough information so the group can provide an answer. Keep in mind that most email lists require you to subscribe before you can post a question.

[2] http://www.nabble.com
[3] http://gmane.org

Many people prefer forums for support. Many OSGIS projects have a forum linked to their website. These can be a valuable source of information and are usually searchable. Here you can find users helping users, as well as information from the project members.

Another great resource is http://gis.stackexchange.com

Sometimes nothing beats real-time support like you can get on IRC. Many projects maintain a presence on IRC. For example, at any one time on irc.freenode.net you might find the following channels: *#grass*, *#postgis*, *#gdal*, *#mapserver*, and *#qgis*. If you don't know what those projects are, never fear. We cover most of these in our survey of OSGIS applications in *Appendix A: Survey of Desktop GIS Software*, on page 299.

IRC has its own unique culture as does each channel. Probably the key thing to remember, apart from doing your homework first, is that people on IRC are almost *always* doing something else at the same time. If you ask a question and nobody answers, it means one of several things. First, nobody is around who knows the answer. People who can't help you often aren't compelled to tell you. Second, the people who do know the answer may be busy at the moment and haven't seen your question yet. And lastly, it's possible your question got lost in the rest of the traffic. Just because no one answers doesn't mean they are snobs, arrogant, or hate you. Your best approach is to hang out for a bit on a channel until you figure out the dynamics.

> **Ask the Right Questions**
> _____
>
> OK, so you need help, and you're ready to ask for it. Nothing will bring on the silence like a poorly asked question. Remember, the people who know the answer are probably pretty busy and have invested a fair amount of time in collecting the knowledge about the application of interest. You need to do the same. Read the documentation, search the Web, and make your best attempt at discovering the answer yourself. You'll learn more from the experience and gain some of that "knowledge."
>
> If you still need help, provide enough information so someone has a reasonable chance of helping you. This typically includes the version of software you are using, your operating system and its version, and exactly what you were trying to do. With most OSGIS applications running on at least three or more platforms, each having its own set of unique issues, this information is pretty important.
>
> Ask the right questions, provide the right information, and you'll get the help you need.

You can also get commercial support for many of the applications discussed in this book. Most OSGIS applications provide information about support on their websites. In addition, a list of support providers is available on the Open Source Geospatial Foundation (OSGeo) website.[4]

4 http://www.osgeo.org/search_profile

Although it has been the subject of some heated debates between the closed and open source groups, most people who have needed support for OSGIS are happy with the experience. If you need support, it's out there and readily available.

4.9 Where to Find Data

By now you realize (or already knew) that without data, we can't do much with OSGIS. For those of you already deeply entrenched in the GIS world, you pretty much know where to search for data; feel free to skip ahead. If you are just getting started with GIS, this is a pretty common question. Your desktop GIS toolkit isn't much good without any data to play with.

The availability of free data depends on where you are in the world. If you are lucky, you live in a country that freely provides data collected by the government. If you are not so lucky, you may have to pay, sometimes quite steeply, to get the data. Don't despair—there is a lot of free data available to get you started.

In reality, there are two types of data: base data and "your" data. Base data is just that—you lay it down as a base for the rest of your map. Examples of base data are country boundaries, rivers, towns, and the DRG that Harrison downloaded in our first encounter with him. Your data is data you have acquired or created for your specific purpose. A simple example is GPS tracks from your latest road trip. You can probably find much of the base data you need for free—let's explore some of the sources of free data.

Clearinghouse Network

One way to find data is to use the Federal Geographic Data Committee's (FGDC) clearinghouse network.[5] The clearinghouse contains nodes (servers) from around the world that contain data and are searchable. Oftentimes you can find the data you need using the clearinghouse search engine.

[5] http://fgdc.gov/

Geo.Data.gov

Another source we mentioned previously is geo.data.gov.[6] This site was established to be "Your One Stop for Finding and Using Geographic Data." Searching for data on geodata.gov yields a list of results containing links to the metadata or website for each dataset. Some of the data may be available for download. In other cases, you'll find that it's available for viewing only through a web map interface using your web browser.

[6] http://geo.data.gov

Other Sources

In the end, the old miner's adage about finding gold applies to geospatial data. Oftentimes the greenhorns would arrive on the gold fields and be clueless. They sought out the sage advice of the old-timers to get them started.

Greenhorn - "Where's the best place to prospect for gold?"

OldTimer - "Gold is where you find it."

There are a lot of sources for data on the Internet, and a bit of judicious searching can lead to good finds. For additional sources to get you started in your data-prospecting adventure, see the list at geospatialdesktop.com.[7]

[7] http://geospatialdesktop.com/data

4.10 Next Step

We've gotten much of the preliminaries out of the way, learned a bit about what OSGIS can do for us, and also looked at some of the things to keep in mind along the way. Now it's time to get into some software and actually do something.

If you want to get the "birds-eye" view of what's available in the open source desktop GIS world, take a look at *Appendix A: Survey of Desktop GIS Software*, on page 299.

Now let's get going and view some data.

5

Working with Vector Data

In this chapter, we'll start working with vector data (points, lines, polygons) by viewing, editing, and analyzing various datasets. Not only will we view data, but we'll look at tweaking the way data is displayed to make it convey more information at a glance.

5.1 Viewing Data

Viewing data is like the "Hello, World!" application that everyone writes when learning a new programming language. It's the first thing you're going to want to do with any GIS application. Let's start out by seeing what kind of things we can do with vector data using open source GIS software. If you recall Harrison's original project, he first just wanted to view bird locations. We'll take a similar approach and start by viewing some sample vector data.

Viewing data is really more like visualizing the relationships between the features. You can get a lot of information by simply viewing features and applying some special rendering techniques.

When it comes to software, we have a lot of choices for viewing GIS data (see *Appendix A: Survey of Desktop GIS Software*). As we begin to explore our data, we'll use several different applications to give you a feel for what's available.

[1] http://geospatialdesktop.
com/sample_data

Before we can begin, we obviously need some data to work with. If you don't have a shapefile or two handy, you can download[1] a sample dataset and use it to follow along.

We will be using this dataset throughout the following chapters when we need to illustrate some basic functions or concepts. The dataset includes world borders, cities, and a nice raster image of the earth (which we'll use in a later chapter when we work with rasters).

What is a Shapefile?

A *shapefile* stores vector features and their attributes. A given shapefile can contain only one type of feature: points, lines, or polygons.

The term is actually a bit misleading, since a shapefile always consists of at least three separate files. For example, a shapefile named alaska would consist of the following:

- alaska.shp containing the spatial features
- alaska.dbf containing the attributes
- alaska.shx, which is an index file that allows random access to features in the alaska.shp file

In addition to the three main files described here, you might also find alaska.sbx, alaska.sbn, and alaska.qix files. These are additional index files used by some applications. One last file you'll often find associated with a shapefile is a prj file. This file contains the projection information for the shapefile, including the geodetic datum.

If you are sharing a shapefile with someone, make sure you include at least the shp, dbf, and shx files; otherwise, it will be unusable.

Choosing a Viewer

Most of the applications in *Appendix A: Survey of Desktop GIS Software*, on page 299, that work with vector data go beyond a viewer. Let's use several of them to look at the sample data. Of course, you don't need to use all of them, but following along will help you decide which is best for you. For help on installing any of the applications, see *Appendix B: Installing Software*, on page 317.

The truth is that nearly all the OSGIS viewers use a similar user interface. If you can use one, you can figure out the others. Let's start by viewing the world borders data using the User Friendly Desktop Internet GIS, uDig.

Simple Viewing

If you need help installing uDig, take a look at Section 17.4 on page 320. OK, let's fire up uDig so we can get a look at that sample data:

In the examples that follow we will be using uDig 1.21

- *Linux*: Change to the udig subdirectory, and run udig.
- *Mac OS X*: Double-click the uDig icon in your Applications folder.
- *Windows*: Click the Start button, find the uDig program folder in Program Files, and choose uDig.

When you first start uDig, you are presented with a start-up screen. You can explore the options, but if you are anxious to get busy, click the curved arrow in the upper right of the workspace. This gets us to the business end of uDig.

The uDig workspace isn't much to look at the first time you run it. You'll notice that when uDig starts up, it displays a fairly typical Tip of the Day dialog box. Feel free to click through the tips and see what pearls of wisdom you can find. You can turn off this feature if it bothers you (or you've read them all).

Now let's load the world borders layer to get a feel for how uDig manages layers, as well as the options for symbolizing features. To view the data from our sample dataset, start by clicking the Layer menu and then choosing Add. This opens the Data Sources dialog box, as shown in Figure 5.1, on the following page.

As you can see from the Data Sources dialog box, uDig supports a good selection of formats. Let's start by adding our shapefile of all the countries in the world. Since this is a file-based data store,

A file-based data store is a fancy way to say a GIS data file on your disk drive as opposed to web-accessible or spatial database data.

Figure 5.1: uDig Data Sources Dialog

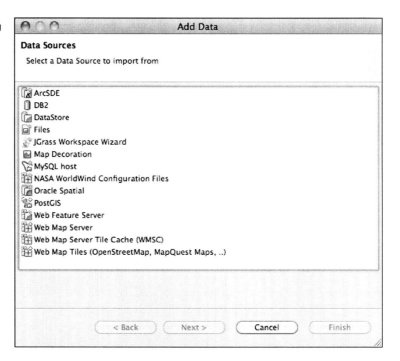

we choose Files from the Data Sources dialog box and click Next, which opens a file selection dialog box from which we can choose our shapefile. We navigate to the directory containing the shape-file (in this case world_borders.shp) and click Open (or whatever the standard dialog box calls it on your platform). This loads the shapefile into uDig and displays it, as shown in Figure 5.2, on the next page.

We've closed the various tabs below the map canvas to maximize the map area and still give you a feel for the interface.

If you are feeling adventurous, go ahead and load the cities layer as well, using the same process.

Moving Around

If you're following along, you should be looking at the countries of the world. Take a look at New Zealand—it's pretty small.   This

In case you're having trouble, New Zealand is located east of Australia on the right side of the map.

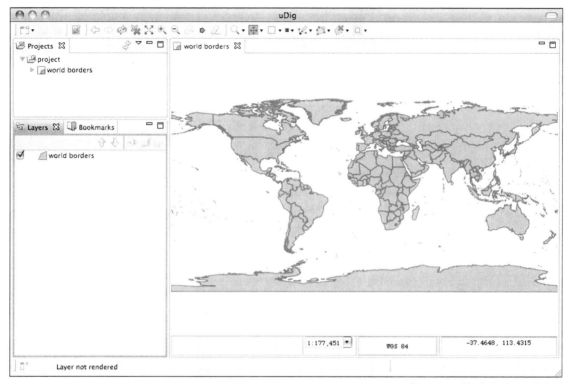

Figure 5.2: uDig displaying world borders

is where navigation tools come into play. Every GIS application, whether it be on the Web or the desktop, has a way to navigate around the map. uDig, of course, supports the usual zoom, pan, and identify functions common to all applications.

Let's get a closer look at New Zealand. Select the Adjust Current Zoom tool from the toolbar. It's the magnifying glass with the dropdown caret next to it. If you are unsure which tool it is, hover the mouse for a few seconds, and you'll get a tooltip to help you out. Find New Zealand, drag a box around it, and then release the mouse. uDig will zoom the view to cover the region of the box. You now have a better view of New Zealand. You can continue to zoom in as much as you like by dragging boxes with the mouse.

So now that we've zoomed into the gnat's eyebrow, we need to determine how to get back out. There are a couple of ways do it. First we can go back to the full view (extent) by using the Zoom to Layers tool in the toolbar above the layer list. This will zoom to include all layers on the map. This may not be what we want if we just want to look at the full view of the world_borders layer. To zoom to just its extent, right-click world_borders in the layer list, and choose Zoom to Layer.

We can also zoom out incrementally by using the Zoom Out tool on the main toolbar. Unlike the Adjust Current Zoom tool, this tool is a one-shot affair—you don't interact with the map when using it. With each click, the map is zoomed out by a fixed amount. By now you've probably noticed its cohort, the Zoom In tool. Clicking it zooms the map in by a fixed amount.

One last way we can navigate the map is by panning. To pan the map, select the Pan Map View tool (actually both this and the Adjust Current Zoom buttons are tool groups but contain only one tool) from the main toolbar, and drag with the mouse to change the view. You can pan all around the map using this method.

By combining the pan and zoom tools, you can pretty much navigate around the world until your heart's content. You can change the map view in other ways, but we'll leave that for you to discover.

5.2   Rendering a Story

Now it's time to change the way the world looks. This is known as *symbolizing* your data, and you can do it in several ways. Are you happy with the colors uDig chose for the layers? If you are like most people, you have preferences when it comes to these things, and my guess is you're going to want to change the way things look. The simplest, of course, is just a single color for all features, and this is in fact the way all vector layers look when first added to uDig. In uDig you can change the outline color, fill, and marker symbols using the Style Editor. The Style Editor also allows you to turn on labeling and set the maximum and minimum scales at which the layer is displayed.

Is It Small or Large Scale?

This can be a constant source of confusion when talking to people about maps, whether they be paper or digital. Let's sort it out now.

A small-scale map covers a large area, whereas a large-scale map covers a smaller area on the ground. The terms *large* and *small* are based on the representative fraction that shows the relationship of one unit on the map to one unit on the ground. A map scale of 1:8,000,000 is smaller than one of 1:24,000 since 1/8,000,000 is a smaller fraction than 1/24,000.

Simple enough. If you ever get confused, just think in terms of fractions, and you'll be able to sort out the small from the large.

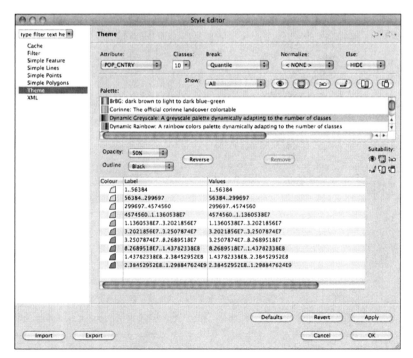

Figure 5.3: Classifying countries by population

Our friend Harrison, being the inquisitive sort, quickly decides he wants to be able to tell at a glance where the most populated countries are in the world. Well, we are in luck. Our world_borders layer just happens to have an attribute named POP_CNTRY that contains the population of each country. To make Harrison happy, we can use what's termed a *class break method* to symbolize the data. The Style Editor has a Theme panel to classify the layer based on an attribute. In Figure 5.3, on the preceding page, we can see the settings we can use to classify the world boundaries based on population. We set up ten classes based on the range of population values. The less populated countries will be lighter in color, while the countries with the largest populations will be rendered in darker shades.

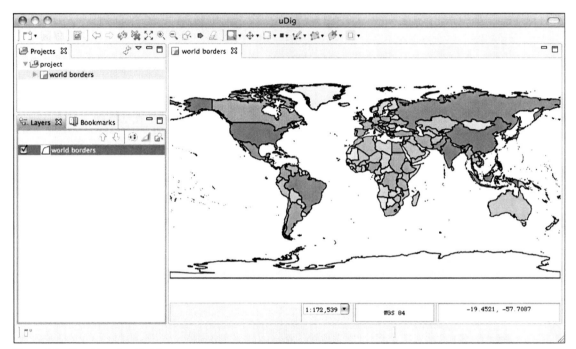

Figure 5.4: Countries classified by population

The results of the classification are shown in Figure 5.4. As you might have suspected, China and India are rendered among the most populous. We could refine our classification to get a finer-grained view of population by changing the number of classes or the method. This is a common way to render data to make it tell a story. Some of the other OSGIS applications we will look later offer even more ways to symbolize your data.

5.3 *Looking at Attribute Data*

In the previous section, you might have wondered how we knew about the POP_CNTRY field in the world_borders shapefile. Well, there are a number of ways to examine the attribute data associated with a layer. One of the quickest ways is with ogrinfo, a utility that is part of the GDAL/OGR suite. We'll take a more detailed look at ogrinfo and friends later in Section 13.2, *Using GDAL and OGR*, on page 212. If ogrinfo is not already on your system, see Section 17.7, *FWTools*, on page 321 for information on installing FWTools.[2] Then you can open a command shell on your system and do the following:

[2] FWTools is a suite of tools that contains many applications, including ogrinfo.

```
ogrinfo -so -al world_borders.shp
INFO: Open of 'world_borders.shp'
      using driver 'ESRI Shapefile' successful.

Layer name: world_borders
Geometry: Polygon
Feature Count: 3784
Extent: (-180.000000, -90.000000) - (180.000000, 83.623596)
Layer SRS WKT:
GEOGCS["WGS 84",
    DATUM["WGS_1984",
        SPHEROID["WGS 84",6378137,298.257223563,
            AUTHORITY["EPSG","7030"]],
        AUTHORITY["EPSG","6326"]],
    PRIMEM["Greenwich",0,
        AUTHORITY["EPSG","8901"]],
    UNIT["degree",0.01745329251994328,
        AUTHORITY["EPSG","9122"]],
    AUTHORITY["EPSG","4326"]]
CAT: Real (16.0)
FIPS_CNTRY: String (80.0)
CNTRY_NAME: String (80.0)
AREA: Real (15.2)
POP_CNTRY: Real (15.2)
```

Near the end of the output you'll find the fields included in the dataset. Note that POP_CNTRY is listed last, and the output indicates that it is a numeric field.

Of course, all the desktop GIS applications provide a way to not only determine which fields are in a dataset but actually view the data itself. In uDig we just click the Table View tab below the map view, and we get a nicely formatted view of the data, as shown in Figure 5.5. We'll examine working with data in other applications in a bit.

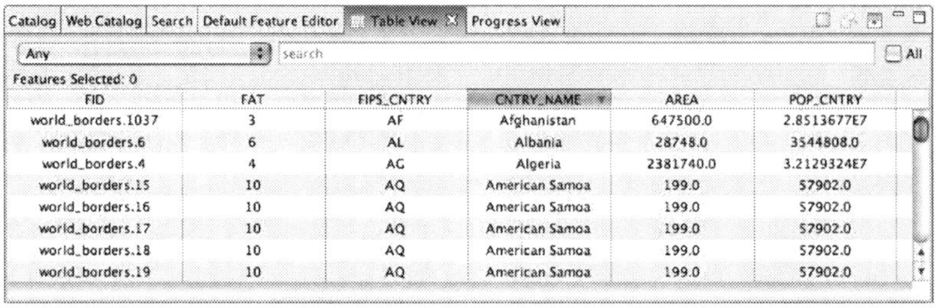
Catalog
Web Catalog
Search
Default Feature Editor
Table View
Progress View
Any
search
All
Features Selected: 0
FID
FAT
FIPS_CNTRY
CNTRY_NAME
AREA
POP_CNTRY
world_borders.1037 3 AF Afghanistan 647500.0 2.8513677E7
world_borders.6 6 AL Albania 28748.0 3544808.0
world_borders.4 4 AG Algeria 2381740.0 3.2129324E7
world_borders.15 10 AQ American Samoa 199.0 57902.0
world_borders.16 10 AQ American Samoa 199.0 57902.0
world_borders.17 10 AQ American Samoa 199.0 57902.0
world_borders.18 10 AQ American Samoa 199.0 57902.0
world_borders.19 10 AQ American Samoa 199.0 57902.0

Figure 5.5: Viewing attributes in uDig

Viewing the attribute table is good for just browsing around. Let's look at some more advanced ways to view and render our data.

5.4 *Advanced Viewing and Rendering*

Harrison has some more bird-sighting data he collected in his travels. He wants to view the sightings in a number of ways, including by species and number of birds per site. This will allow him to quickly identify where he saw individual species and large groups of the same species. Fortunately, there are some advanced rendering techniques that can help him out.

Let's use QGIS to help Harrison classify his data. If you haven't installed QGIS yet, take a look at Section 17.3, *Quantum GIS*, on page 319 for some hints. Then start up QGIS:

- *Linux*: Change to the QGIS install subdirectory, and run QGIS or use the desktop icon if installed on your platform.
- *Mac OS X*: Double-click the QGIS icon in your `Applications` folder.
- *Windows*: Click the Start button, find the QGIS program folder in Program Files, and choose Quantum GIS.

Once QGIS starts up, you are presented with an empty legend and map canvas. In QGIS, functions are accessible from both the menu and the toolbar. Before we get to helping Harrison, let's explore the interface a bit by loading our world_borders and cities layers. Since these are vector layers, find the tool to load a vector or use Add a Vector Layer from the Layer menu. QGIS has a lot of tools on its many toolbars, so it's best to familiarize yourself with them up front. You can do this by hovering the mouse over each tool to view the tooltip or, better yet, by reading the *User Guide*[3] that comes with QGIS. It contains a summary of the tools and includes pictures of the icons to help you get started.

[3] The *User Guide* is distributed with QGIS and accessible from the Help menu.

Once you click the tool or menu option to load a vector layer, the Add vector layer dialog box is displayed. This dialog allows you to select the type of vector data source you want to use. In our case, we want to load a file so make sure the File radio button is selected and click the Browse button to open the file dialog box. Navigate to the directory where you placed the sample data, or you can use your own shapefile data if you have some available. Note you can choose more than one layer from the list by using the Shift or Ctrl key. This allows us to quickly add more than one layer to the map. Select both the world_borders and cities shapefiles, and click the Open button to select them. This returns you to the Add vector layer dialog—click Open to load the layers. If you don't see the shapefiles in the file dialog box, make sure the filter (Files of Type) selector is set to display ESRI Shapefiles.

When loading a shapefile, look for the file with the .shp extension

Once the layers load, you should see something similar to Figure 5.6, on the next page. You might notice something right off, apart from the potentially atrocious colors QGIS has chosen for our two layers. The cities layer is "underneath" the world borders. This is because QGIS isn't very clever in loading layers and didn't know it should put your point data on top of your polygon data. We can easily fix this by dragging the cities layer to the top of the list in the legend. That solves the ordering problem, but the colors are still bothersome. Fortunately, QGIS has a wide range of options for symbolizing layers.

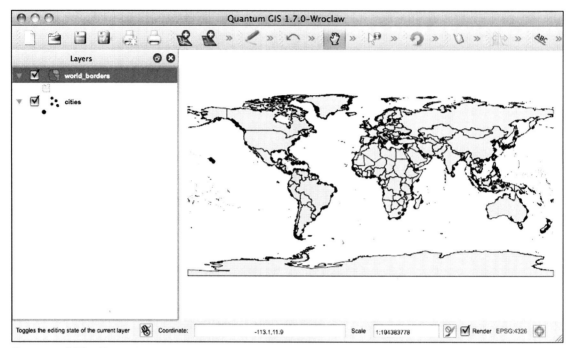

Figure 5.6: QGIS with sample data loaded

Fixing the Appearance

To fix the layers, we will use the Layer Properties dialog box. You can access the dialog box by double-clicking a layer or by right-clicking and choosing Properties from the pop-up menu. Let's start by modifying the look of the world. As you can see in Figure 5.7, on the facing page, there is the somewhat busy Layer Properties dialog box and the current settings for the world. Let's take a bit of time to explore the options available. First off note that the current color of your world_borders layer is selected. We'll change that in a moment. The dialog box is organized into tabs, the default being Style since that is the most often accessed.

For more information on the options available from the Layer Properties dialog, see Section 19.1, *Vector Properties and Symbology Options*.

Let's change that random color of the world_borders to something more pleasing. If you don't have the properties dialog box open, double-click the world_borders layer. Let's make the land a light

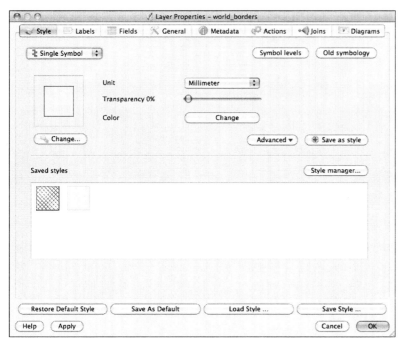

Figure 5.7: QGIS vector layer properties

brown color. To begin, click the Change button under the color box. This brings up the Symbol properties dialog which allows us to not only change the color, but also set things like fill style, borders, and offsets. To set the color, click on the Change button and select a nice brown color from the color selector. We'll do the same with the outline (border color) for the countries, in this case choosing a nice blue color. Once you have done that, click OK to return to the Layer properties dialog. Click Apply to see the changes or OK to accept the changes and close the dialog box.

If you just want to change the fill color, click on the Change button underneath the Transparency slider.

Let's set one last option before we take a look at the results. QGIS allows you to set a background color for the map canvas. Often this can be used to improve the appearance of the map, especially when you have large areas of white space. To set a background color, choose Project Properties from the Settings menu. To set the color, activate the General tab, and then click the color box to the right of the Background Color label. Close the dialog box, and refresh the map. The result of all the changes, as shown in Figure 5.8, shows a nice light blue ocean and the countries of the world neatly delineated.

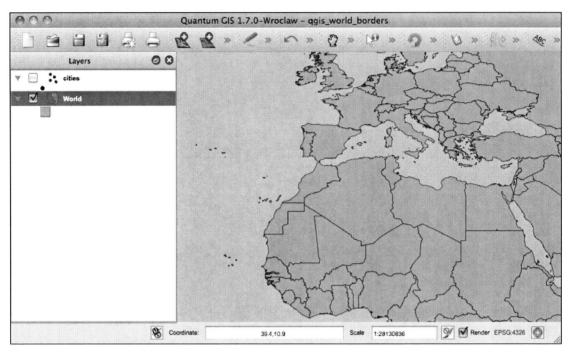

Figure 5.8:   Nicely   rendered
world_borders layer

One other thing to note about the result. We changed the display name of the layer from its arcane world_borders to World. You can set the display name for each of your layers from the General tab of the Layer properties dialog.

Now that we are proficient in adding a layer and adjusting its appearance, let's take a look at Harrison's bird data.

Viewing the Bird Data

To answer all of Harrison's questions about his data, we will use the Continuous Color, Graduated Symbol, and Unique Value renderers in QGIS.

Using Continuous Colors

In continuous color rendering, you set a color for the minimum and maximum values in your data, and it automatically assigns colors to each feature. It turns out this is a quick way to render Harrison's bird sightings to get a feel for the relative distribution of birds. To use continuous color rendering, open the Layer Properties dialog and click on the Old symbology button. The vector Layer Properties dialog box with the continuous color settings for our bird data is is shown in Figure 5.9.

For more about Old and New symbology, see *Style* in *Appendix D: Quantum GIS Basics*, on page 357

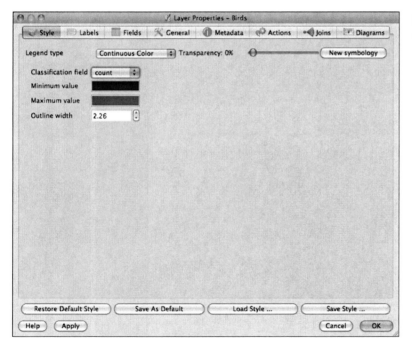

Figure 5.9: QGIS continuous color renderer settings

To set up the renderer, we selected a start and end (minimum and maximum) color. You could use any colors, but we chose a blue to red transition, going from dark blue to dark red. You could just as easily go from an orange to a dark green. The number of birds per site are represented from the fewest (dark blue) to a moderate number (purple) to the most (dark red). The count field from the attribute table contains the number of birds per site and is used to classify the data.

When we apply the renderer, we get a nice display, as shown in Figure 5.10. Notice that we didn't have to specify anything about the data or the individual classes—in fact, there are no options to do that. It gives us a relative view of how the bird counts are distributed, but it is purely qualitative. We can't tell from the legend what a particular color represents in terms of actual number of birds at a given location. Of course, we could use the identify tool (we haven't talked about this yet) to find out.

In this and other examples, you'll notice we've added a background image to enhance the display. You'll see how to add rasters in the next chapter.

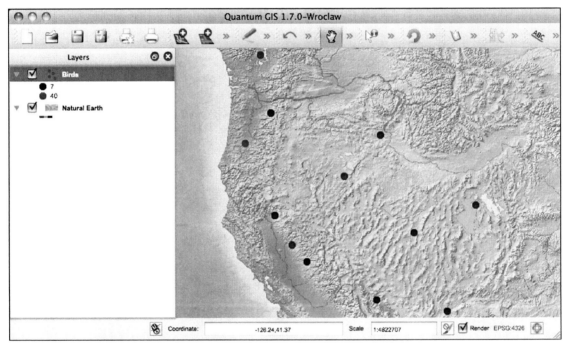

Figure 5.10: QGIS continuous color renderer results

This is a quick way to render the data and get a feel for how things are distributed. Harrison isn't fully satisfied with the result—he wants more control, and as we'll see, the graduated renderer is better suited to the task.

Graduated Symbols

Let's take a quick look at using the graduated symbol render in QGIS. The renderer settings, all ready to go, are shown in Figure 5.11.

Here we're using the "New" symbology for our graduated symbol rendering

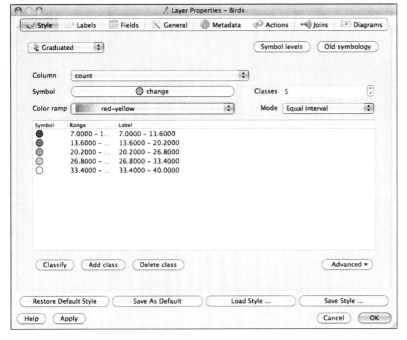

Figure 5.11: QGIS graduated renderer settings

In our settings, we used the red-yellow color ramp to specify that the sites with the lowest bird counts should be a dark red color, while those with the least are rendered in yellow. As a result, you can see a lot of red dots in Figure 5.12, on the following page. This is because we used only five classes. If we wanted a more granular view of the counts, we would need to increase the number of classes. Another way to refine the rendering is to edit the values in the class breaks. QGIS allows you to edit the ranges used for each class by double-clicking the number range in the list of classes (Figure 5.11). You can adjust the ranges for each class to get the result you want.

The graduated renderer gives us (and Harrison) a quick way to spot the locations with the highest bird count, just by looking at the col-

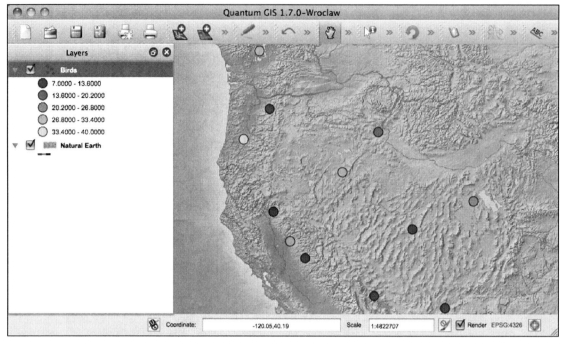

Figure 5.12: QGIS graduated renderer results

ors of the dots on the map. We could create an even more effective display by changing the symbol size of the dots for each class, starting with a smaller point size and increasing to a much larger size for the last class. In this way, the size of the dots conveys the relative number of birds at each site, and you can get a very quick idea of the distribution with just a glance.

Unique Values

Harrison can now get an idea of the bird counts at each site by using the continuous color or graduated renderers. But he also wants to view the information by species. This is where the unique value renderer comes into play.

The unique value renderer is useful for visualizing things that are the same. By that I mean rendering features using the same color when they have the same value for an attribute. Some common examples are the following:

- Displaying all the polygons for a land type in the same color, such as state lands vs. city lands
- Displaying volcanoes by their type
- Displaying roads by type: interstate, highway, secondary road, and primitive road
- And of course displaying birds by name

The common thread in that list is: display "xyz" by *type*. That's the purpose of a unique value renderer.

Figure 5.13: Unique value renderer for birds

To display the bird sightings by name, we set up the unique value renderer as shown in Figure 5.13. We can adjust the colors for each bird by clicking its name in the list and changing the fill color. We can also adjust the style and size of the marker symbols. Once we are happy with the setting and click OK, we get a nice display of

When using New symbology, the Categorized renderer is used to display unique values.

our locations by bird name, as shown in Figure 5.14.

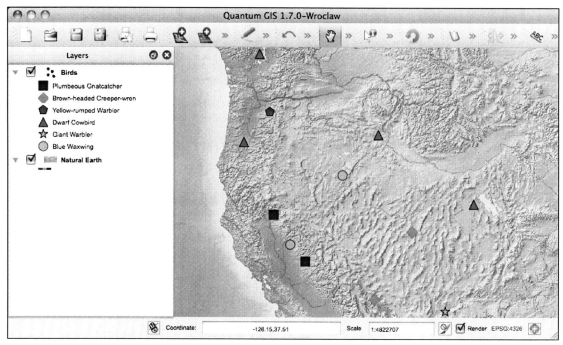

Figure 5.14: Viewing birds by name

Now Harrison is happy, and we have gotten a good look at using renderers to help us understand our data. While we are here, let's look at one more example that has nothing to do with birds. So far in this section we have been working with point locations. Let's take an example that is a bit more colorful and is composed of polygons—a geologic map.

A geologic map portrays rocks by type (note that word *type* again), and each type should be rendered in the same color. The unique renderer setup to display our geologic map is shown in Figure 5.15. Notice that the rendering is done using the UNIT field. This field contains the abbreviation for the rock types.

As with the graduated renderer, you must set the color and style for each unique value. In the case of this geologic map (which is for the

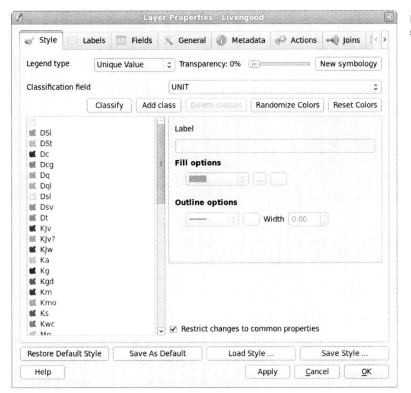

Figure 5.15: QGIS unique renderer settings for a geologic map

Livengood quadrangle in Alaska.[4] The result of our efforts is shown in Figure 5.16, on the following page. A couple of points about the result. First, the colors don't represent those "standard" for geologic units—so you geologists out there don't get too excited. The second point is, if you wanted to make the colors match the standard, you would have to manually tweak the color for each unit.

Now that we have a good idea of how to load and render vector layers to tell a story, let's look under the hood and see what we can do with the attributes associated with our GIS data.

5.5   Making Attribute Data Work for You

So far, we haven't really dealt much with the attributes associated with our data layers, other than using them with the various render-

[4] You can download this and other geologic maps for Alaska from http://pubs.usgs.gov/of/1998/of98-133-a/arc/covers.

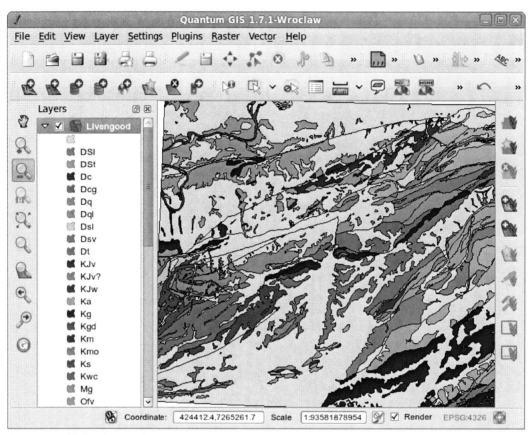

Figure 5.16: QGIS unique renderer
result

ers. In this section, we'll get into working with our attribute data
and making it work for us. We'll leave Harrison's birds alone to
roost for a while and use the cities layer in the sample dataset to
identify features, view the attribute table, select features, and attach
actions to attributes. We'll be using QGIS to illustrate how to work
with attribute data, but you'll likely find similar capabilities in the
other applications we've mentioned so far.

Identifying Features

If you want to follow along, open QGIS and load the cities and
world_borders layers.

Here is one of the most common questions when working with GIS data—what is that feature, and what can I learn about it? It's also one of the simplest operations we can perform. Once we load the cities layer, we have 606 dots on our map. The next trick is to determine which is which. Sure, we can identify some of them just by looking, assuming we are familiar with the country. This is where the Identify tool comes into play—it's used to query a feature on the map. Simply zoom to an area of interest, activate the layer by clicking its name in the legend, click the Identify tool, and click the feature. QGIS will dig up some information about the feature and open the results dialog box, as shown in Figure 5.17.

Figure 5.17: QGIS Identify results

The results show each field name in the left column and the corresponding value in the right column. If more than one feature is returned, you'll have to expand the result nodes to see the field values. In Figure 5.17, we have identified Dublin, Ireland, and can see that it is a capital city with a population of 1,140,000.

If you attempt to identify a feature and QGIS tells you it can't find anything at that location—despite that you're sure you clicked it—you'll have to adjust the search radius used for finding a feature.

This setting is on the Map Tools tab of the Options dialog box, accessible from the Settings menu. It's specified as a percentage of the map width. If the default value isn't working, try increasing it. If you increase it too much, you will end up with multiple features returned instead of the one you want. A default setting of 0.5% is probably a good starting point for a map with global extent. If you still don't get any results, check to make sure you have activated the layer by clicking its name in the legend.

Selecting Features

Selecting features is another common operation when working with attributes. A selection set is used for a number of operations, as we'll see shortly. Here we'll illustrate a simple usage—zooming to the extent of the selection set.

In QGIS there are two ways to select a feature. The first is using the select tools, located next to the Identify button. You have several options for making a selection:

- Select single feature - click on a single feature
- Select features by rectangle - drag a rectangle around the features
- Select features by polygon - construct a polygon around the features
- Select features by freehand - draw a freehand line around the features
- Select features by radius - draw a circle around the features

Choose the method you want to use from the drop-down list, activate the layer, and then make the selection. The selected features will be drawn in a highlighted color (the default is yellow). You can customize the highlight color from the Options dialog box.

The other way to select features is to use the attribute table selection tools which we will look at in Section 5.5, on the facing page.

Once you have a selection set, you can zoom to it by using the Zoom to Selection tool or by using View→Zoom to Selection menu option.

Using the Attribute Table

By now it's clear that GIS layers have attributes associated with the features.[5] The attribute table not only gives us a view into the data behind the features but in a typical application allows us to edit, select, and search.

In QGIS, as in all desktop GIS applications, you can view the attribute table for a layer. The attribute table for the `cities` layer is shown in Figure 5.18. You'll note that there is a row for each feature (city) and columns for each attribute (field) in the layer.

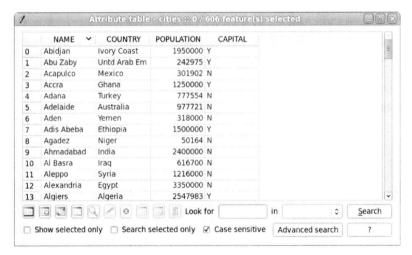

Figure 5.18: Attribute table for the cities layer

This is a very busy dialog box with lots of buttons and features. We'll start with the basics and come back to the more advanced features later. First, you can sort the table by any field by clicking the column header. Clicking repeatedly toggles between ascending and descending sort order. You can scroll through the table and randomly browse the attributes. This is a good way to introduce yourself to a newly acquired dataset.

You can select features by holding down the left mouse button and dragging down through the rows. You can also select rows by using the `Shift` and `Ctrl` keys, just as you do in a selection box on your operating system.

As you select records in the attribute table, the corresponding features on the map canvas are highlighted.

Quick Search

Say we want to find a particular city—for example, Cuzco in Peru. There are several ways we could do it:

- Using the Identify tool, we could randomly click the cities in Peru until we find it.
- Turn on labels for the cities, zoom to Peru, and look around until we find it. This works pretty well, but if we didn't know Cuzco was in Peru, it would become a bit more difficult.
- We can open up the attribute table, sort it by name, and scroll down until we find Cuzco.
- We can search for it.

These methods are in increasing order of efficiency. For small datasets, browsing the attribute table until you find what you are looking for is reasonable. When you get hundreds of records, it can become a chore, and searching becomes the way to go.

You can quickly search the table for a city by entering a search term in the Look for box, selecting the field you want to search from the drop-down list, and clicking the Search button. In our case, we want to enter "Cuzco" as the term to search for, and we want to look in the NAME field. The search will return both full or partial matches. When searching, matching records are selected and the features are highlighted on the map. The view in the attribute table will not change. If you only want to see the matching records, click the Show selected only checkbox.

Several tools at the bottom of the attribute dialog box can be used to manipulate the selection set. Use the mouse to hover over each to learn its function. The tools available are as follows:

- Unselect all
- Move selection to top

- Invert selection
- Copy selected rows to the clipboard
- Zoom map to the selected rows

If you copy the selection to the clipboard and paste into a text editor, you get a comma-delimited list of the attributes in the selected rows, complete with a header row containing the field names. Doing this for Cuzco and pasting it, we get the following:

```
wkt_geom,NAME,COUNTRY,POPULATION,CAPITAL
POINT(-71.860001 -13.600000),Cuzco,Peru,    184550,N
```

The output can be used for importing into another application for further manipulation or reporting purposes. You might be wondering about the funny-looking POINT notation. It's the Well-Known Text (WKT) representation of a point, consisting of the feature type keyword, in this case POINT, and the coordinates (X and Y) separated by a space. The X and Y values are in the coordinate system of the layer. In the case of our cities layer, it's geographic (longitude, latitude). We also have the name, the country, its population, and whether the city is a capital.

The X and Y of Latitude and Longitude

In conversation, most people say "latitude and longitude," not the other way around. In fact, you will most often see it written that way as well. So, it's natural for people to assume the latitude = X and longitude = Y (since we say X,Y), but this isn't so. It's an extremely common mistake, especially for newcomers to the GIS realm. For the record, lines of longitude run vertically and measure units in the X direction. Lines of latitude run horizontally and measure units in the Y direction.

Advanced Search

The attribute table also provides an advanced search query capability where you can really narrow down what you are looking for in the dataset. The quick-search feature allows us to specify only a single search term and field to search. With the advanced query, we

[6] SQL is a standard language for querying and updating a database.

can be more specific using SQL.[6] Don't worry if you aren't a SQL expert or don't even know what it stands for—QGIS makes it easy, as we will see.

Say, for example, we want to find all the cities in the world with more than two million people that are also capital cities. We can easily do this with the Search Query Builder. To access the builder, open the attribute table, and click the Advanced search button. The query builder populated with the terms needed to find the cities of interest is shown in Figure 5.19. Now let's take a look at how this dialog box works.

Figure 5.19: Search Query Builder

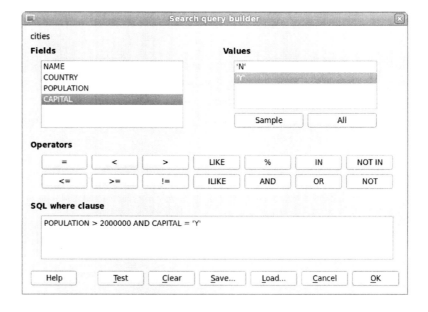

The fields in the attribute table are listed on the left side of the dialog box. On the right is a box that can be used to display the values for a field. This allows you to get a preview of a field's contents to aid you in building the query. To preview a sample of the data, use the Sample button. This will pull a subset of the unique values for the field. Viewing a sample is a good method to use when the attribute table is very large because it saves time. If you want to see all possible values for a field, use the All button.

These two boxes also serve another purpose. You can add an item in the box to the query by double-clicking it. To start building our sample query, we double-click the POPULATION field to add it to the SQL where clause. To complete the population part of the query, we click the "greater than" operator button and then type in "2000000". This gives us the following:

```
POPULATION >  2000000
```

When adding field names and operators, QGIS automatically inserts the needed spaces. If we test the query now using the Test button, it reports 110 matching features (cities). To complete the query, we click the CAPITAL field and click the ALL button to get a list of all possible values. It's easy to see there are two possibilities—Y and N. Since we are using a value from the list, we don't have to type anything to complete the query. First we add the AND operator by clicking the button in the Operators section. To add the final parts, just double-click the CAPITAL field, click the = operator, and double-click Y in the value list. Our query is now complete:

```
POPULATION > 2000000 AND CAPITAL = 'Y'
```

If we test this query, QGIS reports forty-one matching cities. When we click OK, the attribute table will have those forty-one cities selected. We can promote them to the top of the table using the Move Selected to Top button in the Attribute Table toolbar. Since these features are selected, they will be drawn the map using the selection color preference you have set in your QGIS preferences or the project properties.

You may have noticed the Save and Load buttons on the query builder dialog. Once you have created a query that works you can save it to an XML file. Saved queries can be loaded again into the query builder and used to select records. This is handy if you find yourself doing the same queries repeatedly in the course of your work. You can create a single directory to contain your queries, or you can organize them by QGIS project or even datasets.

With the query builder you can construct very complex queries.

If you are familiar with SQL, you can bypass the click-and-build routine and enter the where clause manually. In either case, the ability to define custom searches is a powerful tool. In a future chapter, we will see how to use this same concept to create a subset or multiple views of a layer.

Using Attribute Actions

Let's conclude our look at working with attribute data by doing something useful with the attributes in the `cities` layer. QGIS has the concept of *attribute actions*, in other words, performing some action (think task) using the value of an attribute. With an attribute action, you can call another application and pass the value of the attribute to it. Here are some potential uses for this feature:

- Do a web search based on one or more attribute values.
- Display a photo based on a location stored as a layer attribute.
- Submit values to a URL that creates a report.
- Query a database based on attribute values.

There really is no limit to the way in which you can use actions to integrate QGIS with other applications and tools. Let's take a simple example and do a Google search of a city using the results from the `Identify` tool. Attribute actions are defined from the Actions tab on the vector Layer Properties dialog box. The steps to create an action are as follows:

1. Determine the attribute field(s) needed.
2. Determine the application that will "drive" the action.
3. Construct the argument string using the attribute value(s).
4. Create the action.

Creating the Action

In our example, we are going to use the name of the city and pass it to Google. The application we need is a web browser. In this case, we'll use Firefox. The URL to search for something with Google is `http://google.com/search?q=find_me`, where `find_me` is

the search term. Attribute actions work by replacing specific strings
in our argument list with the values from the attribute table. The
format of these strings is simple—it's a percent sign (%) followed
by the name of the field we want to use. In our example, we therefore need to use %NAME as the replaceable parameter. Putting this
altogether gives us the following action:

```
firefox http://google.com/search?q="%NAME"
```

One important thing to note here—the browser must be in our path.
If not, the action will fail. The alternative and perhaps the safe way
is to fully specify the path to the browser:

```
/usr/bin/firefox http://google.com/search?q="%NAME"
```

If you have spaces or other oddities you need to specify in the action, use double quotes around the entire thing. They will be ig-
nored when the action is executed but safely allow you to specify
the path and other exotic parameters. Since city names have spaces
in them, we need to quote either the entire URL or the %NAME parameter.

So now that we have the text for the action figured out, let's put it
all together to create and use the action. First open the properties
dialog box for the cities layer and click the Actions tab. Give the
action a name, "Google search" will work, and then enter the text of
the action. If you are following along, make sure to adjust the path
for your browser. Click the Insert action button to add the action to
the list. We could go ahead and create other named actions in the
same way. If you make a mistake, you can click the action in the list
and edit it.

When you are finished editing, click the Update action button to
save your changes. Notice we didn't use the Insert field button
and its associated drop-down list of field names. That's because
we determined beforehand we were using %NAME. Also notice the
browse button to the right of the action text box. If you click this
button, it lets you browse to the location of the application you want
to use to execute the action. In our example, we could have used

it to browse to the location of Firefox (`/usr/bin/firefox`). This is useful if you aren't sure of the full path for the application needed to execute the action.

To complete the creation of the action, click OK. It's now ready to use.

Using the Action

Now that we have it, let's see how to use it. The results of identifying a city on the map are shown in Figure 5.20. If you compare this with Figure 5.17, on page 61, you will see we have expanded the (Actions) node to reveal our action, appropriately labeled "Google search." To execute an action, you can click it or right-click and choose the action from the pop-up list (if we had defined more than one action, we would choose it from this list). When clicked, QGIS will launch Firefox and execute the Google search for New Orleans. You can identify another city and use the action to perform a Google search on it. Depending on your operating system and browser, it may reuse the current browser window or open a new one.

Figure 5.20: Attribute action enabled in QGIS

Now it should be clear where attribute actions could come in handy. There is one last trick you can use when defining actions. If you use the special parameter %% instead of a field name, QGIS will replace it with the value of the currently highlighted field in the identify results list or the attribute table. In the case of our example, this would allow us to do a search on any field value in the layer. Most of our fields in the `cities` layer aren't well suited for Google searching, but the `COUNTRY` field would return useful results. Being able to specify values in this way, as well as the ability to define multiple actions gives us a lot of flexibility. See the QGIS User Manual for more information.[7]

[7] http://qgis.org/en/documentation/manuals.html

What you do with attribute actions is now limited only by your imagination and cleverness.

Now that we've exercised our vector muscles, let's move on and work with some raster data.

6

Working with Raster Data

Raster data is everywhere in the GIS world. You can use it as a background layer for your vector data or do full-blown analysis with it. In this chapter, our goal is to get you up and running with raster data. In later chapters, we'll delve into some analysis and manipulation.

Nearly every OSGIS desktop application can display at least some raster formats—and some more than others. In particular, those applications that are based on the GDAL library can support an impressive range of raster data. You can find a partial list of formats that GDAL supports in Section 16.2, *GDAL/OGR*, on page 309. Both QGIS and GRASS use the GDAL library for reading and writing raster data.

6.1 Viewing Raster Data

We'll start with something simple in our endeavor and load a TIFF image. A fairly common thing you might want to do is view a topographic map of your area. You might recall that this is what Harrison used as a background for his bird data. In the United States, many of these rasters can be downloaded from the U.S. Geological Survey website. Using a topographic map as a base is useful when you want to view your vector data (for example, GPS tracks

[1] http://www.archive.org/details/maps_usgs

[2] In case you're interested, we grabbed http://www.archive.org/details/usgs_drg_mt_47113_g1.

and waypoints) over it.

Let's download a TIFF from the Internet Archive of USGS Maps.[1] You can pick any state you like—for our example, we'll grab a random image from Montana.[2] You can pick one for your area by browsing the archive by state. We'll need both the `tif` and `tfw` files. Once you have your raster and the world file, you can view it in QGIS.

The QGIS toolbar contains a button for loading rasters (it's right next to the Vector button), or you can choose `Add Raster Layer` from the `Layer` menu. Rasters in QGIS are loaded using the standard file dialog box—of course this will vary in appearance depending on your operating system. To select a TIFF, we need to make sure the filter is set properly. The QGIS raster dialog box has filters for a number of data types, including GeoTIFF, ERDAS Imagine, and USGS Digital Elevation Models. To get started, just make sure the GeoTIFF is selected in the Files of Type drop-down box, and navigate to the location of the TIFF file. Select it and click the Open button. QGIS opens and displays the image at full extent. In Figure 6.1, on the facing page, we have loaded the raster and zoomed to the northeast corner of the Montana DRG.

What Is a Georeferenced TIFF?

It's a TIFF image that has metadata (information) about its coordinate system. This information can be associated with the image in a couple of ways. In a GeoTIFF, the coordinate system information is "embedded" in the file itself—in other words, it is self-contained.

The other way to specify coordinate information is with a world file. A world file contains information about the map units per pixel in the image, as well as the real-world coordinates of the upper-left corner. For a TIFF, the world file usually has a `tfw` extension. World files are used with other georeferenced image formats, including JPEG (`jgw`) and PNG (`pgw`). For software that employs the GDAL library for raster access, the `wld` file can be used with TIFF, JPEG, PNG, and other supported formats.

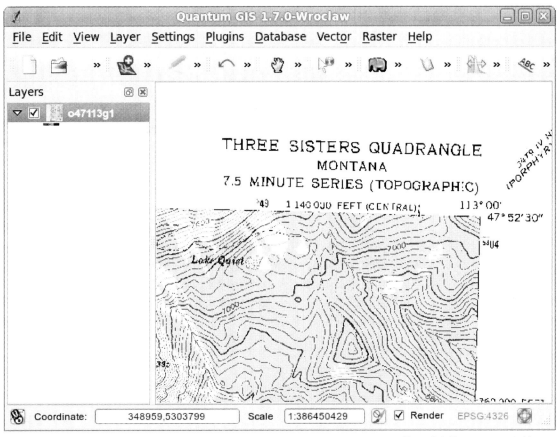

Figure 6.1: Montana topographic map in QGIS

The first thing you're likely to notice is that the raster has a bunch of text annotations around the border. We call this the *collar*, and although it has good information, it's a bit distracting when we're working with our data. This becomes readily apparent when you download the DRG adjacent to it and want to display them in a seamless fashion. We'll take up this issue in a bit and show you how to create a seamless raster from several TIFF images in Section 12.6, *Clipping Rasters with GRASS*, on page 190.

With your topographic map loaded, feel free to zoom around and explore the countryside. If you take the Identify tool and click the raster, you'll find it doesn't yield much in the way of information.

In fact, the only thing it will tell you is the palette index of the pixel where you click. This is because rasters are composed of cells (a pixel is a cell) and contain only one value. In the case of a DRG, that's the palette index number. For each index, there is a corresponding color value. So for the Montana DRG, if we click a lake or stream, we find that the palette index is 2. This isn't all that useful, but when we get to Section 6.3, *Intelligent Rasters*, on page 83 you'll see other rasters where the cell values convey significant information.

The last thing we need to mention is the coordinate system for this raster. If you open the Raster properties dialog box (just double-click the raster name in the legend) and click the Metadata tab, you'll find that my Montana raster is in UTM Zone 12, NAD27 datum. You can glean that information from the Layer Spatial Reference System section of the dialog box. Another way to get the same information is with gdalinfo, a utility that comes with GDAL and is included in FWTools:[3]

[3] If it's not already on your system, see Section 17.7 for information on installing FWTools.

```
gdalinfo o47113g1.tif
Driver: GTiff/GeoTIFF
Size is 4769, 6920
Coordinate System is:
PROJCS["NAD27 / UTM zone 12N",
    GEOGCS["NAD27",
        DATUM["North_American_Datum_1927",
            SPHEROID["Clarke 1866",6378206.4,294.9786982139006,
                AUTHORITY["EPSG","7008"]],
            AUTHORITY["EPSG","6267"]],
        PRIMEM["Greenwich",0],
        UNIT["degree",0.0174532925199433],
        AUTHORITY["EPSG","4267"]],
    PROJECTION["Transverse_Mercator"],
    PARAMETER["latitude_of_origin",0],
    PARAMETER["central_meridian",-111],
    PARAMETER["scale_factor",0.9996],
    PARAMETER["false_easting",500000],
    PARAMETER["false_northing",0],
    UNIT["metre",1,
        AUTHORITY["EPSG","9001"]],
    AUTHORITY["EPSG","26712"]]
Origin = (339751.702040000003763,5305307.515394000336528)
Pixel Size = (2.438400000000000,-2.438400000000000)
...
```

We didn't include all the output from gdalinfo; we included just
enough for you to see the projection information. If you recall earlier, I told you that when downloading, we needed the tfw file,
which is actually the world file for the raster. The fact that the
projection information is reported by gdalinfo means that it is a
GeoTIFF and contains not only the coordinate information needed
to properly display it but also the projection information. You don't
need the world file for a GeoTIFF. The fact that the DRG was a
GeoTIFF wasn't readily apparent from the website, so we played it
safe and downloaded the world file as well. Fortunately, we didn't
waste much bandwidth downloading it since world files are only a
few hundred bytes in size.

Let's return to our global theme now and view a raster mosaic, cour-
tesy of NASA Visible Earth.[4] If you want to follow along, you can
fetch the raster from http://geospatialdesktop.com/sample_data
or the NASA site. Loading it into QGIS gives us a world mosaic, as
shown in Figure 6.2, on the next page.

The NASA raster is a georeferenced image in geographic coordinates, meaning it can be used in conjunction with our world vector
layers. If you look carefully at Figure 6.2, you'll notice we've added
the cities on top of the raster, just to prove they line up. Fortunately,
both the raster and vector data have geographic coordinates in the
same *datum*. In case you're wondering, a datum is a model of the
shape of the earth used to measure positions. In this case, the coordinates are in WGS 84, the same datum commonly used in modern
GPS units. We'll take a further look at datums and projections in
Chapter 11, *Projections and Coordinate Systems*, on page 159.

Raster Properties

Now we'll take a brief look at some of the properties associated
with a raster. Just like vector layers, in QGIS there is a properties
dialog box that allows us to adjust the appearance of the image, as
well as get some information about it. You can access the properties
dialog box by double-clicking the layer name or right-clicking it and
choosing Properties. If this sounds familiar, you are correct. It's

[4] The ev11612_land_ocean_ice
8192 image is owned and provided by NASA. The image was
obtained from the Visible Earth
(http://visibleearth.nasa.gov)
and developed by the Earth
Observatory team (http:
//earthobservatory.nasa.gov)

Figure 6.2:   NASA world mosaic
viewed in QGIS

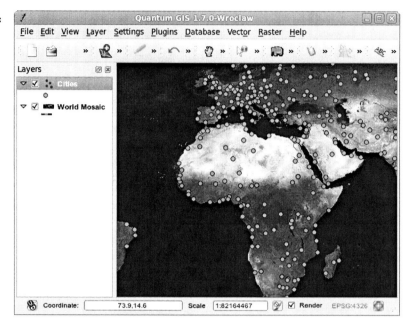

the same method used for accessing the vector properties dialog
box. Figure 6.3, on the next page, shows the raster properties dialog
box.

As you can see, the raster properties dialog box bears a bit of resemblance to the vector properties dialog box. They both have Style,
General, and Metadata tabs. Rather than look at each of these in
detail, we'll focus on a couple of things you need to know to effectively use images in QGIS. On the Symbology tab, you have the
choice to display the image as either color or grayscale. By default,
the appropriate display mode is chosen for an image when you load
it. In the case of our example, we have a multiband image consisting
of three bands: red, green, and blue. We can do a number of things
with the bands, including changing which band is used for which
color. In other words, we can swap the mapping of color to band
and observe the effects. We can also invert the colors, which swaps
the light and dark colors. The transparency of a raster can be set
using the slider on the Transparency tab, allowing the layers under-

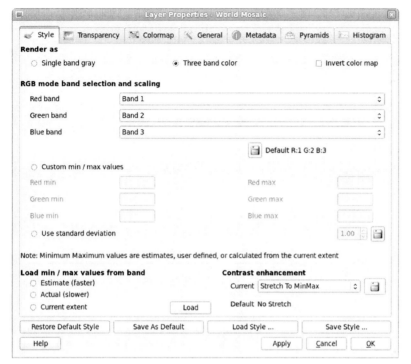

Figure 6.3: QGIS raster properties dialog box

neath to become visible. This can be used to get some interesting effects, such as those in Figure 6.4, on the following page.

The Transparency tab allows you to set custom transparency options by setting either the band to use for transparency or by adding a list of transparent RGB pixel values.

If you are working with a grayscale image, you can adjust the color map and standard deviation used to display the raster. See the QGIS user guide to learn more about these options. One other thing to note is you can display a color image as grayscale by selecting the Grayscale radio button and then setting the band in the Gray drop-down box. So, for example, we can display our color image as grayscale using the green color band (band 2) to determine the appearance of the image. This might be used to bring out features or characteristics not visible when the image is displayed in color.

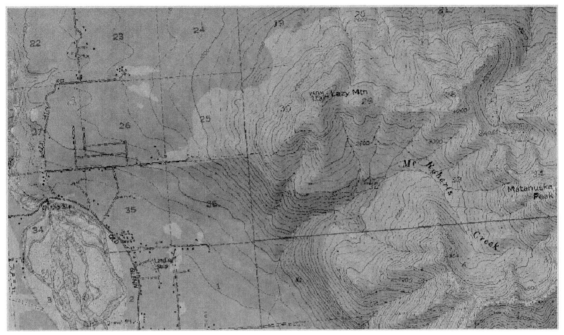

Figure 6.4: Semitransparent digital elevation model draped over a DRG

6.2   Improving Rendering with Pyramids

Pyramids are essentially multiple views of a raster at reduced resolutions. Using pyramids means that the software, in our case QGIS,
doesn't have to draw every detail of the image to get it on the screen.
This is appropriate because all that detail is lost on you at small
scales. QGIS supports building and using pyramids through the
GDAL library. Some software stores pyramids in external files and
others store them within the image itself. This means your original
image will be altered and in fact grow in size when building pyramids. You may want to make a copy of the image before creating
pyramids because the process is not reversible. GDAL supports the
creation of external pyramid files for some raster drivers. If you
don't want to modify your original image, make sure the *Build pyramids internally if possible* checkbox is cleared. In our case, we will use
the default and build external pyramids

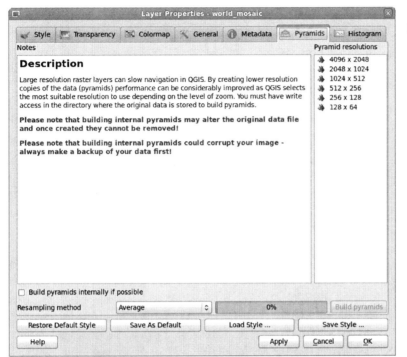

Figure 6.5: QGIS raster pyramids dialog box

To create pyramids, open the raster properties dialog box, and select the Pyramids tab. Note the warning about altering and possibly corrupting the image, and make the backup copy before you proceed. QGIS populates the Pyramid Resolutions list with a set of resolutions appropriate for your image. The dimensions of the world mosaic image are 8,192 by 4,096 pixels. In Figure 6.5, you can see that QGIS offers to build a range of pyramids based on the dimensions of the image. These levels are calculated by dividing the width and height down to some minimum level. Notice the small X over the pyramid icon for each level. This indicates that pyramids do not exist for that level. This of course changes once we build pyramids for the raster. You can choose a sampling method for calculating the pyramids. If you are unsure what to do, just take the default value because it will give you acceptable results. Select the levels you want to build by clicking each one. The more you select, the longer it will take to build, and more important, the larger your image will grow if you choose to build internal pyramids. Click the Build pyramids button, and wait while QGIS generates the pyramids for each selected level in the list. This can take a while, especially for large images. Once complete, the small X will disappear from each level for which pyramids were built.

You should now see improved performance when drawing the image at smaller scales. You might also notice some degradation in the appearance of the image; however, at large scales (when zoomed in), the full quality of the image is preserved. You may have to experiment a bit to determine how many and which levels you want to build. For this reason, be sure to keep a copy of the original image in a safe place.

You can also create pyramids using the GDAL utility gdaladdo. By default the pyramids are added to the original raster—to store them externally use the -ro switch. If you want to create pyramids for a lot or rasters, using gdaladdo is the way to go. You can write a small script (shell, Python, Ruby, or Perl) to process each file in a directory and add the pyramids. Here is an example of using gdaladdo to create pyramids for our Montana DRG:

```
$ ls -lh o47113g1*
-rwxr-xr-x 1 gsherman gsherman 6.8M 2011-11-02 16:42 o47113g1.tif
$ gdaladdo -ro -r average o47113g1.tif 2 4 8 16
0...10...20...30...40...50...60...70...80...90...100 - done.
$ ls -lh o47113g1*
-rwxr-xr-x 1 gsherman gsherman 6.8M 2011-11-02 16:42 o47113g1.tif
-rw-r--r-- 1 gsherman gsherman  12M 2011-11-02 16:45 o47113g1.tif.ovr
```

Here we created four levels of pyramids using the "average" resampling algorithm. We listed the size before and after the operation. Notice that building pyramids added an external overview file o47113g1.tif.ovr that is nearly twice the size of the original. Without the -ro switch the pyramids would have been stored in o47113g1.tif, increasing its size to near 12 megabytes. For more information on gdaladdo and its options, see the documentation.[5]

Creating pyramids for your raster data can give you a huge performance gain when rendering at various scales. You may want

[5] http://www.gdal.org/gdaladdo.html

to make a backup copy and then experiment with the various levels and sampling methods to see which provide the best results for your data.

6.3   Intelligent Rasters

All rasters are intelligent—at least in the sense that they convey information. So far we have looked at rasters (DRGs and our world mosaic) that are useful for general viewing or as a background layer. Typically these rasters have cell values that don't really mean anything—they are just a value used to determine how each pixel should be colored. Cell values in a raster (or *grid* as these are often called) can be either integer or floating point, depending on the capabilities of your software. You might be wondering what sort of data can be represented by a raster. The answer is anything you can count, measure, or identify for a defined area on the ground. Some examples include the following:

- Rock or soil types, specified by a unique numeric code for each
- Vegetation types
- Elevations
- Quantity of an element such as gold or silver, determined by sampling and analysis

Why not use a vector layer rather than raster to delineate data by "type"? Oftentimes it's appropriate to use a vector layer, as we did for our geologic map in Figure 5.16, on page 60. Here each rock type is represented by a polygon and rendered by value. It really depends on what you need to accomplish with the data. In the case of a grid containing quantities that we want to use in an analysis, the raster model is the right choice.

Let's look at a couple of examples of what we might term a "smarter raster," in this case a Digital Elevation Model (DEM) and a grid containing measured quantities of silver.

Digital Elevation Models

The cells in a DEM contain an elevation value. For any location on the DEM, we can determine the elevation by examining the value of the cell. The smaller the size of the cells, the more resolution you will get from the DEM. We can do a number of interesting things with a DEM, including the following:

- Display it as is for a backdrop.
- Create a shaded relief to display terrain.
- Perform arithmetic operations on the individual cells to create new values.
- Create contours from the elevations.

We'll look at DEMs in more depth in Chapter 12, *Geoprocessing*, on page 171, where we'll see how to use them in analyzing, creating contours, and generating shaded reliefs—also known as *hillshades*.

The Grid of Silver

Let's see how we can use a grid to do some qualitative analysis. Harrison's interests are diverse—this time he's off to do some prospecting for silver. In terms of the methodology, he could just as easily be examining the concentration of hazardous materials or four-leaf clovers.

Harrison has found the results of a soil sampling survey that was done over a regular grid. Each sample was analyzed for a number of elements, but he is interested only in silver. Fortunately for Harrison, along with the results is a raster grid in Arc/Info Binary Grid format. Harrison fires up QGIS and finds he can easily load the grid by choosing the `Add a Raster Layer` tool from the toolbar and changing the file type to "Arc/Info Binary Grid". The grid consists of a directory containing several files. The one we (and Harrison) are interested in is the `adf` file. This contains the grid data, and QGIS knows how to display this format. In Figure 6.6, on the facing page, you can see what the grid looks like loaded into QGIS and underlain by the DRG for the region. In order to see the underlying DRG, we've made the grid partially transparent using the transparency slider on the raster properties dialog box.

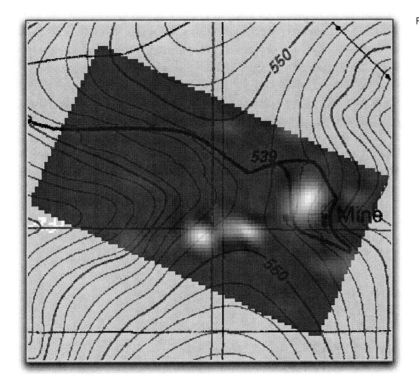

Figure 6.6: Grid of silver values

Notice that the grid is "tilted." That's because it has been transformed from the original coordinate system, a simple X-Y grid, to one that places it where it belongs in the real world. From looking at it, we can conclude that whoever laid out the grid didn't do it using cardinal directions of north-south and east-west. Transforming it to the proper coordinate system also gives it a ragged appearance along the edges since it's not possible to approximate a straight line given the cell size in this grid.

We can use the Identify tool just as we did when working with our vector data (Section 5.5, *Identifying Features*, on page 60). Unlike vector data where we might have several attributes for a feature, what we're identifying here is the value for a single cell. You'll also notice that although for a vector layer we can use the select tools and open the attribute table, we can't do it for the grid. The only information we can gain is the cell value.

To help Harrison with his qualitative analysis, let's identify a few cells to get a feel for the data. The dark cells to the west (left side of the map) are low values,[6] typically in the neighborhood of 0.2. If we look at the brighter areas of the grid, we find values up to 3.65 or greater. So, just by looking at the grid we can get a feel for where the higher values are, once we know the pattern.

[6] The units on this grid are in parts per million (ppm). Nobody is going to get rich off the silver in this grid.

Using the Metadata tab in the raster properties dialog box we can get some interesting information about the grid. If you scroll to the bottom of the dialog box, you'll find some statistics of interest:

Property	Value
Band	Undefined
Band No	1
Min Val	-0.2643643320
Max Val	3.6971910000
Range	3.9615553319
Mean	0.3253868880
Sum of Squares	991.8232408759
Standard Deviation	0.4478514579
Sum of All Cells	1609.3635482636
Cell Count	4946

This tells us the minimum and maximum values in the grid, the number of cells, the range, and some other statistics. This is probably a better way to get a quick overview of the distribution of a grid, as opposed to randomly identifying cells.

To make the display a bit more dramatic, we can open the raster properties dialog box and change the color map from grayscale to pseudocolor. When we apply this change, we get the result shown in Figure 6.7, on the facing page. Now the high value areas are red and the low values are blue, making it even easier to visually analyze the distribution of the values.

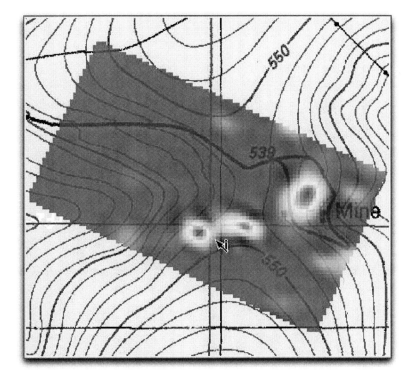

Figure 6.7: Grid of silver values in pseudocolor

We could take our analysis further by using GRASS to create a contour map of the cell values or by generating a hillshade to enhance the appearance of the map. We'll delve into contouring and creating hillshades, as well as some other raster manipulation, when we get to Chapter 12, *Geoprocessing*, on page 171.

Now that we've learned a bit about rasters, let's take a look at digitizing some vector data using our OSGIS software.

7

Digitizing and Editing Vector Data

One of the strengths of desktop GIS is the ability to create new data. Although your favorite desktop application can be a data consumer, it can also be a creator. In this chapter, we'll look at creating vector data and some of the reasons why you might want to do so.

7.1 Simple Digitizing

If you remember Harrison's original bird project, one of the things he wanted to do was create a new vector layer for the lakes in one of his birding areas (see Figure 3.3, on page 20). Once he had the lakes in a new vector layer, he could do some more advanced GIS processing to create a buffer and test his hypothesis regarding birds and nearness to lakes. This is a pretty simple digitizing project, one that we'll do for Harrison.

Picking a Tool

As you've gathered, we could use a bunch of tools to digitize the lakes. Since we're going for simple here, either uDig or QGIS is a good choice. You could use OpenJUMP or GRASS as well. Since we are going to do some geoprocessing with this layer (a fancy way of saying create a buffer), it makes sense to create it in a flexible format that we can import into whatever application we choose. The obvious choice is a shapefile, although we could just as easily have chosen PostGIS—but since we won't talk about that until Chapter 9, *Spatial Databases*, on page 111, we'll keep it simple.

To build Harrison's layer, we'll use QGIS and create a shapefile containing lakes as polygons.

Digitizing the Lakes

Figure 7.1: Creating a new shapefile in QGIS

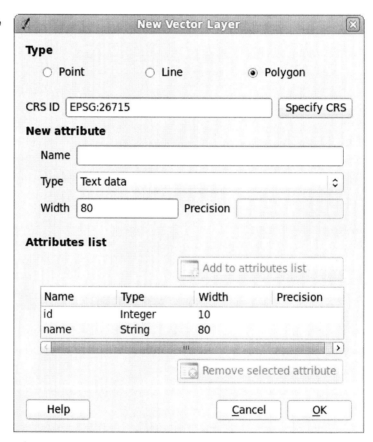

The first step is to fire up QGIS and add the raster we want to digitize from to create the lakes. We're using o48092d8.tiff, a DRG from the Daley Bay Quadrangle in Minnesota.[1] Once we have the raster loaded, we need to create a new vector layer for the lakes.

[1] This DRG is available at http://www.archive.org/download/usgs_drg_mn_48092_d8/o48092d8.tif.

QGIS supports the creation of new shapefiles for editing, but you could just as easily create a new PostgreSQL layer using SQL and edit it. For now we'll create a shapefile with an id field and a name field. To do this, choose Layer→New→New Vector Layer from the menu. In Figure 7.1, on the facing page, you can see the completed layer information with the fields defined. We also entered the CRS to match our DRG so everything will line up properly. By opening the raster properties dialog for the DRG and looking on the General tab we were able to determine the coordinate reference system was EPSG:26715.

Once we click the OK button, QGIS opens the dialog box to save the file. This allows you to navigate to the directory where you want the shapefile to live and give it an appropriate name. Having done that, our shapefile is created and displayed in QGIS as shown in Figure 7.2, on the next page. Of course, there is nothing in it yet.

Now we are ready to digitize. First we need to get the new layer into edit mode by right-clicking it in the legend and choosing Toggle editing from the pop-up menu. A small pencil icon will appear next to the layer name in the legend to indicate that we are now in edit mode. Before we start editing, we need to make sure the digitizing toolbar is visible. If not, right-click in the toolbar area of QGIS, and choose Digitizing from the pop-up menu. In Figure 7.2, on the following page, the digitizing toolbar is visible (it's the one with the pencil icon). You'll notice that some of the buttons are disabled (grayed out) because they operate on a selection set and we don't have one at present. In addition, the digitizing tools are clever—they won't show you tools for feature types not appropriate for your layer—in this case point and line tools. QGIS knows we are editing a polygon layer and won't allow you to create the wrong feature type. Let's start digitizing by clicking the Capture Polygon tool.

First we will digitize the boundary of largest lake in our map view (it's the big one to the west). An important thing to remember when digitizing is to choose an appropriate scale for creating features. To get a gross approximation of the lake, we can just zoom in until we

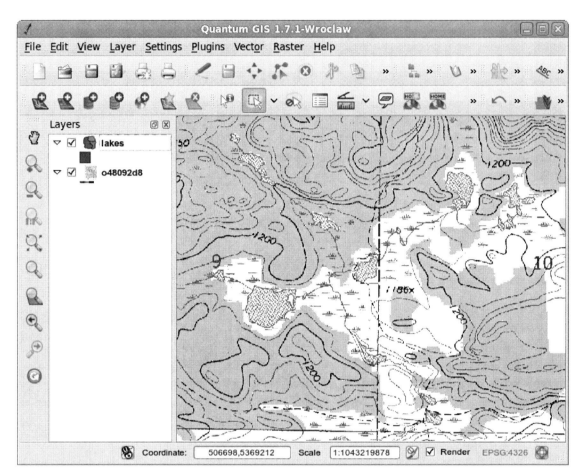

Figure 7.2: QGIS with new layer ready
to edit

see only the lake. If we want a detailed representation of the lake,
we need to zoom in much closer and pan around the map as we
digitize. We'll take a middle-of-the-road approach here and zoom
in somewhat to illustrate navigating the map while digitizing.

To start, we zoom in until the lake pretty much fills the map view. To
digitize the lake, click with the left mouse button and begin moving
along the shoreline, clicking at each location where there is a change
in direction. If you get to the edge of the map canvas and need
to pan to continue digitizing, don't use the pan tool; instead hold

down the Spacebar and move the mouse to pan. If you are zoomed in really close, this technique allows you to work your way around the lake. If you find out that you need to change the zoom level to effectively digitize, you can zoom in and out using the mouse wheel. Both of these navigation techniques are discussed in Section 19.3, *Map Navigation and Bookmarks*, on page 369.

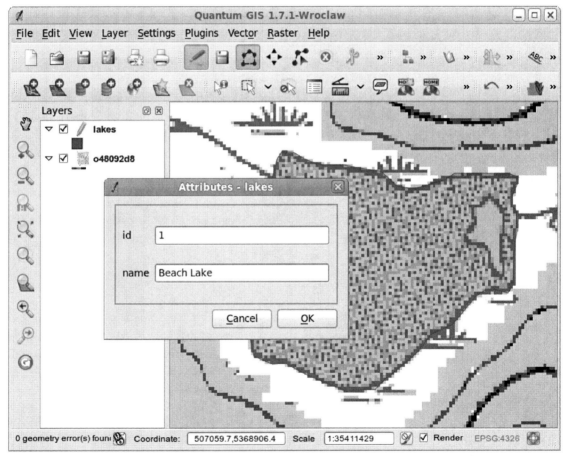

Figure 7.3: Entering attributes for a feature

You'll notice as you digitize around the lake, the display is updated with the polygon as you create it. To complete the lake, right-click at the final point. This opens the attribute dialog box where we enter the attribute values for the feature. In Figure 7.3, we have assigned an ID of 1 and entered "Beach Lake" as the name. When we click OK, these values will be saved for storage in the dbf of our shapefile. That completes Beach Lake; however, the feature hasn't actually been saved yet. To actually write our changes out, we need to stop editing by clicking the Toggle editing tool in the toolbar or right-clicking the layer and choosing Toggle editing. A confirmation dialog box appears, and we have to click the Save button to save the edits. Once we do that, the lake appears like a regular polygon feature, properly colored, and we can both identify it and view the attribute table.

There is a problem though. Beach Lake has an island near its eastern shore. Fortunately the Advanced Digitizing toolbar has an Add Ring tool that will allow us to "cut out" the area of the island. To do this, we simply get back into editing mode, choose the Add Ring tool, and then digitize the island. Right-click to finish it and the island is created. There are no attributes to assign since it isn't part of the lake.

In Figure 7.4, on the next page, you can see the results of our digitizing effort, with the lakes labeled using the name field in the attribute table.

Fixing Mistakes

As you digitize, you are bound to make mistakes. You might find that you didn't follow the shore quite right, made a line too straight when it should have been curved, or just totally bungled the boundary. Typically the best approach is to continue and then correct the problems after you have completed the feature.

All our OSGIS desktop applications that support editing allow you to make adjustments to features by moving, deleting, and inserting vertices. Once you've completed a feature, the vertices are displayed (in QGIS they look like X's). Using the vertex tools on the editing toolbar, you can adjust the boundary to correct any errors.

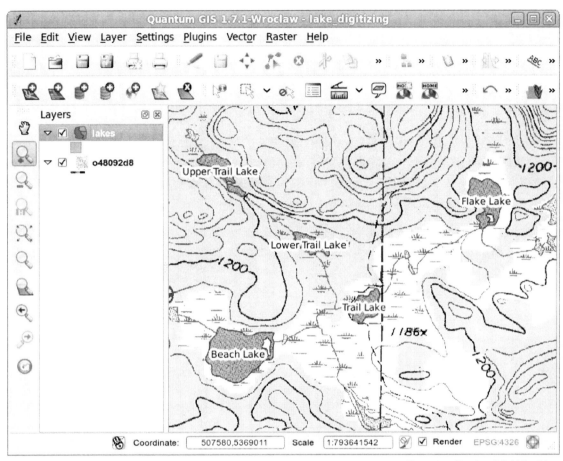

Figure 7.4: Results of digitizing lakes in QGIS

Keeping Your Data Safe

When editing your data with any GIS application, it's a good idea to make sure you have a backup copy. Let's face it, disaster can strike, whether it be a program crash, power outage, or beverage incident. Keeping a current backup of your critical data is just good practice.

Likewise, it's a good idea to save often while editing to minimize any potential loss of your well earned efforts.

Harrison decides he also needs to digitize the streams connecting his lakes. Along the way, he makes a few mistakes, which we'll help him fix. In Figure 7.5, you can see part of Harrison's fist attempt at connecting the lakes using QGIS and a line shapefile for the streams.

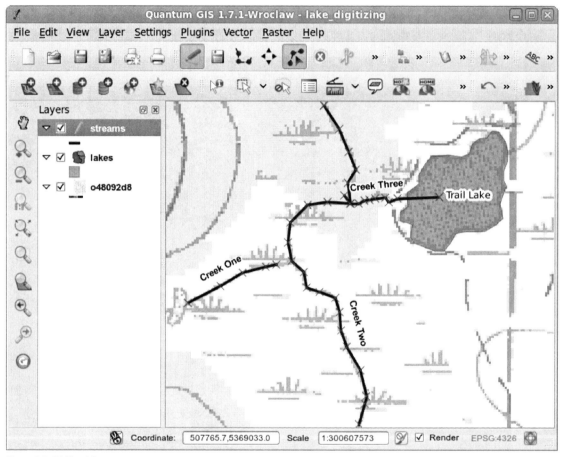

Figure 7.5: Digitized Streams

If you look closely at the streams (we've made them quite wide so they stand out), you'll notice a number of problems. For one, we have a stream that doesn't connect with its neighbor (Creek One). In the case of Creek Three, the line segment on the west overshoots the intersection with Creek Two. These undershoots and overshoots are called *dangles*. Lastly, the eastern end of Creek Three runs too far into the lake on the east. The other thing you'll notice is that Harrison was a bit sloppy in following the stream course, especially approaching the lake to the east.

Figure 7.6: Problems with the digitized streams

Let's use the vertex-editing tools to clean things up a bit. All the errors can be easily corrected just by inserting new vertices where needed and moving existing vertices to intersect where they should. To do this, we first need to set the snapping tolerance. This controls how QGIS snaps to existing vertices when editing. By setting a reasonable tolerance, we can make QGIS "jump" to the closest vertex, thereby making our job easier as well as making sure the line segments actually touch. It's harder than you may think to manually move a vertex and get it exactly on the line.

To determine a proper snapping tolerance, you need to take a look at your data and get an idea of distance between vertices in your lines. We'll set the tolerance in map units, so you may find you need to experiment to get it set right. If you specify too big of a tolerance, QGIS may snap to the wrong vertex, especially if you are dealing with a large number of vertices in close proximity. Set it too small, and it won't find anything. To set the snap tolerance, choose Settings→Snapping Options... from the menu. Remember the tolerance is in map units. If you want to make sure, use the Measure Line tool to examine the distance between vertices to make an educated guess at a proper value. Since our data is in meters, a value of 2.0 seems to work well. This of course depends not only on the map units but also the scale at which you are digitizing. Fortunately, you can tweak the tolerance to get something that works.

Once the tolerance is set, we can move the vertices to fix the dangles. With the Node Tool selected, place the cursor over the vertex to be moved, click on it, and drag it to the new location. When you release the mouse, the vertex is moved, and the shape of the feature changes. You'll notice that when you get close to a vertex, the snapping will kick in and pop the vertex you are moving right on top of the other. You may find that there is no vertex at the location you where your lines should meet. To get the stream to join the other line segment in the right location, make sure the snapping mode (Settings→Snapping Options...) is set to vertex and segment.

To clean up the sloppy work and make the streams match properly, we can add more vertices to solve the problem. First use the Node Tool to add the vertices where needed by double-clicking on the line segment that needs to be modified. Don't worry if you don't get them exact. Once you've added the vertices you need, use the Node Tool to adjust the positions to reflect the path of the stream— or you can add and move each as you go along. If you didn't add enough vertices to accurately portray the feature—no problem—just add some more and move them around until the job is done.

You may find that you have too many vertices or just plain put one where it doesn't belong. In that case, use the Node Tool and click on the node you want to delete. Then press the delete key on your keyboard to remove it. When you delete a vertex, the feature

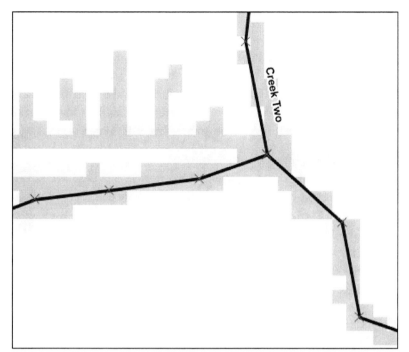

Figure 7.7: Digitized streams with corrections

reshapes itself automatically. In Figure 7.7, you can see the result of fixing a dangle and reshaping the stream to better match with the DRG. As you can see, it's pretty simple to correct mistakes in your data as you go and alter it to make it more precise when your requirements dictate.

7.2   Editing Attribute Data

Now that we have digitized the lakes and streams, everything is in good shape—except we have some problems with attributes associated with the features. If you look at Figure 7.4, on page 95, you'll notice the lake in the northeast corner of the map is named Flake Lake. This is a typo—the actual name is supposed to be Fluke Lake. We can correct this error using the editing capabilities built in to the attribute table.

To fix the name of the lake, we simply open the attribute table for our lakes layer and click the *Toggle editing mode* button (we could also have clicked `Toggle Editing` on the Digitizing Toolbar). Once we do that, any of the items in the attribute table can be modified by clicking them and changing the value. We change the name of the lake to Fluke Lake like it's supposed to be and hit `Enter` to make the change stick. Once we're done editing the table, toggle out of edit mode and click the Save button to save our changes.

It was pretty easy to find the record that needed editing, since we had only five lakes in our layer. What happens if we have 5,000? In that case, you have two options. You can use the search tool, as described in Section 5.5, *Using the Attribute Table*, on page 63, to find the desired record, or you can select the lake on the map canvas using the select tool and then float it to the top of the table using the *Move selection to top* button—or just click the *Show selected only* checkbox.

If you find that you need to make changes to a lot (or all) records in a table you can use the field calculator which is enabled in edit mode. If you are a SQL guru (or want to be) and need to do a lot of data manipulation, then you might consider using something other than a shapefile. A spatial database may be more suited to your needs. We'll take a look at them in Chapter 9, *Spatial Databases*, on page 111.

7.3   *More Digitizing and Editing*

So far we've used simple digitizing and editing techniques to capture some data from Harrison's background DRG. Needless to say, there are more advanced means of digitizing and editing available to us, particularly with GRASS.

We'll look at some additional digitizing tasks in Chapter 10, *Creating Data*, on page 137. Nowadays it's often best to spend a little bit of time looking around for the data you need before you dive into a digitizing project. Oftentimes the data already exists, and you can save a lot of time by just grabbing it from the Internet and moving forward. But at times you are faced with having to create your own data. The methods we have looked at so far form a good foundation for you to launch your own projects.

Speaking of data, let's move on to a look at data formats and what you need to know when working with OSGIS desktop software.

8

Data Formats

One of the challenges in working with GIS software, whether it be proprietary or open source, is making sense of the many data formats you encounter. Let's take a look at some of the common formats you will encounter so you can get an idea of what's out there. We'll also look at where these data formats come from, some of the conversion options, and lastly how to choose a standard data format for your mapping projects.

8.1 Common Formats

So far, we've indirectly discussed a number of formats, including shapefiles, GeoTIFF, grids, PostGIS, and GRASS vector and raster. If you are a casual or intermediate user of OSGIS software, odds are you are going to be using only a few data formats on a regular basis. Typically this means working with the following:

- Shapefiles (shp)
- GeoTIFF or TIFF with world files (tif, tfw)
- JPEG with world files (jpg, jpw)
- GPS data (gpx)

In fact, these are pretty common, and you can accomplish a lot with just these formats. Other vector formats you might run across during your search for data include the following:

- ArcInfo Binary Coverage
- ArcInfo Interchange File (e00)
- MapInfo (.tab, mid, mif)
- SDTS, a data transfer standard for both vector and raster data
- Topologically Integrated Geographic Encoding and Referencing (TIGER) data, used and distributed by the U.S. Census Bureau
- Digital Line Graphics (DLG)

There are a lot of raster formats you might encounter, including the following:

- ERMapper Compressed Wavelets (ecw)
- Erdas Imagine (img)
- Digital Elevation Models (dem)
- JPEG 2000 (jp2, j2k)
- Multi-Resolution Seamless Image Database (MrSID) (sid)
- GTOPO30, a global digital elevation model (DEM) derived from a number of raster and vector sources

Some of the formats are open, meaning that they have a published specification and you can use it to write applications and utilities that work with the format. Others are closed, requiring you to use the vendor-provided API. Of course, this is a concern only if you want to write your own applications and utilities. If you are content with using the OSGIS applications available, someone else has done the hard work for you.

Although it's not important to understand these formats to use them, it does help to know a bit about them so you can determine whether your favorite OSGIS software supports the format. In case you haven't realized it yet, the GDAL/OGR library supports a huge range of vector and raster formats—see Section 16.2, *GDAL/OGR*, on page 309 for lists of additional vector and raster formats you might encounter.

The good news is if you are using OSGIS software that uses GDAL/OGR for accessing raster and vector data (such as GRASS or QGIS), then you have access to most, if not all, of the formats listed.

In Section 13.2, *Using GDAL and OGR*, on page 212, we'll look at using the GDAL/OGR utilities to get information about our data as well as convert and transform both raster and vector layers.

Web-Deliverable Data

Another "format" you'll encounter is data deliverable over the Web. This category of data is often referred to as W*S. The moniker W*S is attached to standards for delivering geospatial data over the Web and includes Web Mapping Service (WMS), Web Features Service (WFS), and Web Coverage Service (WCS). A good chunk of the web mapping applications you might use in your browser get some or all of their data from a W*S service.

Many of our desktop applications include support for at least WMS. This allows us to include data from across the Internet in our mapping projects. The good thing is you don't have to understand the standard or how it works; you just use it and get good data for free.

If you want to get real technical information on WMS, WFS, and WCS, you can find standards on the Open Geospatial Consortium website.[1]

[1] http://www.opengeospatial.org

8.2   Choosing a Standard Format

You might be wondering about a standard format for all your projects. This isn't strictly necessary, although it might make your life easier. Assuming your OSGIS software can handle a multitude of formats, there may be no reason to convert.

Reasons to Standardize

There are some valid reasons to standardize on a data format for your raster and vector data. In reality, you'll probably still have some data that's not in your standard format. Let's take a look at some reasons why you might want to convert to a standard format.

Data Management

GIS data can have several unsavory characteristics—and we're not talking about accuracy or quality. As you begin to work with data, transform it, analyze it, and so forth, you'll find the following:

- It propagates rapidly.
- It grows and hides in places you never expect.
- Unchecked, it rapidly becomes unmanageable.

[2] http://spatialgalaxy.net/gis-data-is-an-illicit-drug/

Some (OK, me) have gone so far as to call GIS data an illicit drug.[2]

You may be wondering how converting to a standard format will improve data management. Well, it's not a silver bullet, but it can aid in creating a logical structure for storing your data. For example, if all your vector data is in shapefiles and you can create a nicely organized directory structure, be it by theme or by project, then your ability to find and use the data you need increases.

Some find that data management is improved by using a spatial database, although there are perhaps better reasons to use one. By storing your vector and/or raster data in a spatial database, you provide one point of entry for all your data needs. There is no question of which server or directory you need to search to find the data you need. We'll look at spatial databases in more detail in Chapter 9, *Spatial Databases*, on page 111.

Another example is GRASS, which uses its own format for storing both raster and vector data. Data in GRASS is organized by location and mapset, making it easy to structure your data collection in a way that can be more easily managed.

Is improved data management alone a reason for converting to a particular format? Probably not, especially if you're talking about converting between file-based formats such as shapefiles and Geo-TIFFs. The chief considerations are making the data discoverable, accessible, and usable. If your formats of choice include file-based data, you should create a structured logical directory layout and naming convention and adhere to it. This will make managing your data much easier and prevent the multiplication of duplicate data sets. Another important management tool is metadata that documents each of your datasets. If you want to take the formal approach and create metadata in a format that others will understand, use the standard.[3] At the very least, you should include a text file describing the data, its origin, and the processing steps used to create it. Fortunately, the metadata standard includes all these components, so you may find it's worth using.

[3] http://fgdc.gov/metadata

Improved Functionality

If you want to do more than display and edit spatial data, then conversion to gain improved functionality is certainly a worthwhile consideration. Harrison's simple analysis of bird sightings using a buffer (Figure 3.3, on page 20) requires something with geoprocessing capability. In this case, Harrison could use QGIS or perhaps convert his digitized lakes (which likely began life as a shapefile) into a PostGIS or GRASS layer. All of these applications give him the functionality he needs to create a buffer.

PostGIS is a good example of a reason to convert. Not only does it improve data management by giving you a "portal" to your data, but it has been certified as OGC compliant and provides the spatial functions specified in the Simple Features Specification for SQL. This means that not only can we display and edit PostGIS data in an application-like QGIS, we also get a whole batch of spatial functions that we can use to query the relationships between features, transform between projections, and create new features. If you find that your work requires more than just simple viewing and editing, then conversion is worth considering.

When software is certified as OGC compliant, you can be assured that it adheres to established standards and can interoperate with other compliant software.

Enhanced GIS Capabilities

The other reason to convert is to gain enhanced GIS functionality. You're probably asking what the difference is between this and the improved functionality aspect we just covered. You can view it as a progression to more powerful and perhaps complex GIS opera-

Geoprocessing tools in QGIS are improving all the time.

tions. Although QGIS and others have a good array of geoprocessing tools, GRASS has the most impressive assortment of vector- and raster-processing tools. Since GRASS stores data in its own format, we need to convert our existing data in order to take advantage of the tools.

Examples of the type of capabilities we're talking about include the following:

- Line-of-sight analysis
- Union and intersection of layers to create a new layer
- Merging raster data
- Mathematical operations on grids
- Contouring

We'll dive into some of these topics later in Chapter 12, *Geoprocessing*, on page 171.

8.3   Conversion Options

So, you decide to convert some or all your data to a new format. The next question is, what tools are available to do the job? Fortunately, in the OSGIS world, conversion between formats is not only commonplace but easy as well.

If you choose to migrate all your data to PostGIS or GRASS, it's not a problem. Both provide the routines to import your data and export it should the need arise.

GRASS Conversion

GRASS provides both vector and raster import/export functions for a nice range of formats. To give you an idea of the capabilities, here is a partial list of the import commands and formats supported by GRASS:

r.in.arc
   Converts an ESRI ARC/INFO ASCII raster file (GRID) into a (binary) raster map layer

r.in.ascii

Converts ASCII raster file to binary raster map layer

r.in.aster

Imports, georeferences, and rectifies an ASTER image

r.in.gdal

Imports a GDAL-supported raster file into a binary raster map layer

r.in.srtm

Imports Shuttle Radar Topography Mission (SRTM) hgt files into GRASS

r.in.wms

Downloads and imports data from WMS servers

v.in.ascii

Creates a vector map from ASCII points file or ASCII vector file

v.in.db

Creates new vector map (point layer) from database table containing coordinates

v.in.dxf

Converts AutoCad DXF files to GRASS format

v.in.e00

Imports an ArcInfo export file e00 to GRASS format

v.in.garmin

Downloads waypoints, routes, and tracks from a Garmin GPS receiver into a vector map

v.in.gpsbabel

Downloads waypoints, routes, and tracks from a GPS receiver or a GPS ASCII file into a vector map using formats supported by gpsbabel

v.in.ogr

Converts OGR-supported formats into a GRASS vector map

You can see from the list of commands that there are a lot of options for getting your data into GRASS. In fact, we didn't list all of them for you, just some of the major ones. For exporting data out of GRASS, there are also a lot of options. We won't list them here, but in case you're curious, the commands are all of the form `r.out.*` for rasters and `v.out.*` for vectors.

PostGIS Conversion

If you choose PostGIS, it supports the loading of data using SQL and the importing/exporting of shapefiles using `shp2pgsql` and `pgsql2shp`. We'll take a look at these two utilities in Section 13.4, *PostGIS*, on page 232. If your source data isn't in shapefile format, you can still import it—you just need a little extra power. In this case, you need to use the Swiss Army knife of conversion tools.

The Swiss Army Knife

Just as you wouldn't go out into the wilderness without your Swiss Army knife (or maybe bug spray), venturing into the world of data conversion without the GDAL/OGR utilities is not advised. These utilities provide conversion between file-based vector and raster formats, as well as spatial databases.

We take an in-depth look at these tools in Chapter 13, *Using Command-Line Tools*, on page 199. For now just keep these commands in the back of your mind: `ogr2ogr`, `gdal_translate`, and `gdalwarp`.

9

Spatial Databases

9.1   Introduction

In this chapter, we take a look at spatial databases. A spatial database allows us to store features, display them, or perform geoprocessing and analysis through a rich set of spatial functions. Some of the advantages of storing data in a spatial database are as follows:

- Attributes and geometry of features are stored together.
- Spatial indexing makes drawing faster at larger scales.
- Spatial queries provide the ability to explore features and their relationships.
- You get better data management.

Structure of a Spatial Database

A spatial database is nothing more than a regular database with support for geometry data types. It typically contains functions to manipulate the geometries and perform spatial queries.

In a spatial database, a table represents a layer, a row is a feature, and a spatial column contains the geometry of the feature.

Open source spatial databases come come in two varieties: server and serverless. Let's deal with the server type first.

What is a Spatial Query?

A spatial query is one that involves features and their relationship to one another. For example, assuming we had the appropriate data in our database, we might ask "Give me the names of all the coffee shops within 10 kilometers of my house." A spatial database is well suited to that type of query and can easily answer that question. Another simple example is finding all the eagle nests within a drainage basin. Of course, you can do much more complex things with spatial queries including transforming and creating new data, as well as projecting data on the fly.

9.2   Open Source Spatial Databases

In the OSGIS world there are currently two options for spatially enabled databases for your server: PostgreSQL[1] with PostGIS[2] and MySQL.[3] Of the two, PostgreSQL/PostGIS is the most mature and feature rich. Originally MySQL implemented many of the OGC spatial functions, but not all of them according to the specification. As of MySQL 5.6.1 true support for testing spatial relationships between geometries has been added. If you want to use the database to do spatial processing and queries, PostgreSQL with PostGIS is the best choice.

Comparison of Open Source Spatial Databases

Both PostGIS and MySQL implement the Open GIS Consortium's OpenGIS Simple Features Specification for SQL. You can find the specification on the OGC website.[4]

The standard is fully implemented in PostgreSQL/PostGIS and the PostGIS implementation has been certified by the OGC as compliant with version 1.1 of the standard. What does this mean to you? It means that PostGIS provides a complete and robust implementation of the standard, along with additional features not in the specification. Since PostGIS has been around longer than the MySQL spatial implementation, more desktop and web mapping clients/servers

[1] http://postgresql.org
[2] http://postgis.refractions.net
[3] http://mysql.org

[4] http://www.opengeospatial.org

fully support it.

9.3 Getting Started with PostGIS

In this section, we'll look at how to enable PostGIS in your Post-greSQL database and load some data from shapefiles and perhaps other sources. We assume you already have a working PostgreSQL install. If not, refer to the installation section of the manual[5] where you will find detailed instructions for getting PostgreSQL up and running on your platform. If you're lucky, PostgreSQL is already installed, and you are ready to proceed with getting PostGIS set up.

[5] http://postgresql.org

Getting PostGIS can be easy if you are running the right platform. You may find a binary version available for download.[6] Otherwise, you will have to build from source. If you are running Linux, be sure to check for a PostGIS binary using your package management tool. Many distributions include PostgreSQL and PostGIS, making it easy to get started. If you are running Windows, the latest PostgreSQL installers include an option to install PostGIS. It's not selected by default, so make sure to you include it when choosing options during the install. Once you have the software in place, you're ready to set up a database and add the PostGIS extension and tables.

[6] http://postgis.refractions.net

Creating a PostGIS-Enabled Database

Since PostGIS is an extension to PostgreSQL, you have to add it to a database in order to use the geometry types and functions. Let's look at a session that creates a new database and enables PostGIS:

```
gsherman@ubuntu:~$ createdb -E UTF8 geospatial_desktop
gsherman@ubuntu:~$ createlang plpgsql geospatial_desktop
gsherman@ubuntu:~$ psql geospatial_desktop
psql (8.4.8)
Type "help" for help.

geospatial_desktop=# \i /usr/share/postgresql/8.4/contrib/postgis-1.5/postgis.sql
SET
BEGIN
CREATE FUNCTION
...
geospatial_desktop=# \i /usr/share/postgresql/8.4/contrib/postgis-1.5/spatial_ref_sys.sql
```

```
INSERT 0 1
...
COMMIT
ANALYZE
geospatial_desktop=# \d
               List of relations
 Schema |       Name        | Type  |  Owner
--------+-------------------+-------+----------
 public | geography_columns | view  | gsherman
 public | geometry_columns  | table | gsherman
 public | spatial_ref_sys   | table | gsherman
(3 rows)
geospatial_desktop=#
```

Let's breakdown what we did and explain the steps. First we created a database using the createdb command, specifying an encoding type of UTF8 (Unicode) using the -E switch. Using Unicode for
the database encoding provides us with the most flexible solution,
especially when storing non-ASCII data. Once the database is created, we add the PostgreSQL procedural language (PL/pgSQL) to
the database using the createlang command and the plpgsql keyword. PostGIS needs PL/pgSQL in order to implement its spatial
types and functions.

Now that we have a database set up and properly configured, the
next step is to load the PostGIS extension into our database. The
commands to do this are provided with PostGIS in the postgis.sql
file. We simply execute this SQL in our newly created database.
There are a number of ways to do this (for example, from a database
client tool such as PgAdminIII); however, we chose to use the PostgreSQL interactive terminal psql.

In psql, we use the \i command to read the file from disk and
execute the SQL statements. In our example, we had to know where
the scripts were installed and specified the full path. You could of
course change to the directory containing postgis.sql and avoid
having to enter the full path. This creates the types and functions.
At this point we have a PostGIS-enabled database, but we aren't
done yet.

PostGIS and Templates

There is an easier way to create additional PostGIS-enabled databases. When PostgreSQL creates a database, it does it by copying an existing database. Usually this is the standard system database `template1`. Anything in the template database ends up in your newly created database. We can use this capability to create a PostGIS template database that can be used when creating a new PostGIS-enabled database.

To create a template, simply create an empty database from the standard template using **createdb -E UTF8 postgis_template** from the command line. Then follow the example in this chapter to load the `postgis.sql` and `spatial_ref_sys.sql` scripts. Once you have the template, use **createdb -E UTF8 -T postgis_template myNewDb** to create a new PostGIS-enabled database.

If you installed PostgreSQL on Windows with the PostGIS option, it should have created a `postgis_template` database for you. In this case, you are ready to start creating your own PostGIS-enabled databases.

The final step is to create the spatial references table that contains more than 2,600 coordinate systems. To do this, we executed the `spatial_ref_sys.sql` file, also provided with PostGIS. Our database is now ready to use for PostGIS data. Using the \d command in `psql` gives us a list of the tables in our new database. In addition to the `spatial_ref_sys` table, you'll notice the `geometry_columns` table and the `geography_columns` view. Let's look at each in a bit more detail.

The geometry_columns Table

The `geometry_columns` table describes the spatially enabled tables in your database. Many applications rely on the records in `geometry_columns` to determine which tables are spatial tables. This table, as well as `spatial_ref_sys`, is described in the OpenGIS Simple Features Specification for SQL.[7] We can view the structure of the `geometry_columns` table using the \d command in the `psql` interac-

[7] http://www.opengeospatial.org

tive terminal:

```
desktop_data=# \d geometry_columns
          Table "public.geometry_columns"
      Column       |         Type          | Modifiers
-------------------+-----------------------+-----------
 f_table_catalog   | character varying(256) | not null
 f_table_schema    | character varying(256) | not null
 f_table_name      | character varying(256) | not null
 f_geometry_column | character varying(256) | not null
 coord_dimension   | integer               | not null
 srid              | integer               | not null
 type              | character varying(30) | not null
Indexes:
  "geometry_columns_pk" PRIMARY KEY, btree (f_table_catalog, f_table_schema,
      f_table_name, f_geometry_column)
```

The first three columns provide a fully qualified name for a table. Some databases may use "catalog", and since the geometry_columns table is a standard, it is included in every OGC-compliant implementation. PostgreSQL doesn't use the concept of a "catalog", so this column will always be blank. By default, all tables in PostgreSQL are placed in the "public" schema. So for a PostGIS-enabled table, we will find the f_table_schema and f_table_name populated with "public" and the table name, respectively.

To avoid problems when upgrading, you shouldn't put your data in the public schema

The f_geometry_column contains the name of the column in your spatial table that contains the geometry. The coord_dimension contains the dimension of the features (2, 3, or 4). The srid is the spatial reference ID column and contains a number that is related to the srid column in the spatial_ref_sys table. This defines the coordinate system for the table. The type field contains information about the feature type contained in the table and contains a keyword such as POINT, LINESTRING, POLYGON, and the MULTI forms of each feature.

With this information, a client application (in other words, your desktop GIS) can quickly determine which spatially enabled tables are available and also collect the information needed to load, tranform, and display the features in a table. The geometry_columns table can be populated in several ways:

- When loading data using shp2pgsql, QGIS, or ogr2ogr, a record is inserted into the geometry_columns table.
- Using the AddGeometryColumn function on a table that does not already have a spatial column. This creates the column in the table and also inserts a record into geometry_columns.
- Manually inserting a record using a SQL insert statement.

Typically you use the AddGeometryColumn function when you create a new table and want to add a geometry column. In this case, you create the table using SQL *without* the geometry column and then use the function to both add the column and create an entry in the geometry_columns table.

In this example we create a new schema (hazards) and store the table in it.

```
geospatial_desktop-# CREATE SCHEMA hazards;
geospatial_desktop-# SET search_path TO hazards,public;
geospatial_desktop-# SHOW search_path;
   search_path
-----------------
 hazards, public
(1 row)
geospatial_desktop=# CREATE TABLE hazards.lakes(LAKE_ID int4,
  LAKE_NAME varchar(32), LAKE_DEPTH float);
CREATE TABLE
geospatial_desktop=# \d
              List of relations
 Schema  |        Name        | Type  |  Owner
---------+--------------------+-------+----------
 hazards | lakes              | table | gsherman
 public  | geography_columns  | view  | gsherman
 public  | geometry_columns   | table | gsherman
 public  | spatial_ref_sys    | table | gsherman
(6 rows)

geospatial_desktop=# select AddGeometryColumn('hazards', 'lakes',
  'the_geom', 4326, 'POLYGON', 2);
                  addgeometrycolumn
------------------------------------------------------
 hazards.lakes.the_geom SRID:4326 TYPE:POLYGON DIMS:2
(1 row)
```

A final note on the geometry_columns table: Some software such as QGIS can search your PostgreSQL database and determine which tables are spatially enabled. To maintain maximum flexibility in your database, you should probably ensure that each spatial table has an entry in the `geometry_columns` table.

The geography_columns View

The `geography_columns` is a view that supports the relatively new PostGIS *geography* data type. This data type is specifically tailored for use with coordinates specified in latitude and longitude. The geographic data type only supports the WGS 84 CRS (SRID 4326) so it is typically suited for work on a global or near global scale.

To create a table that uses the geographic data type:

```
CREATE TABLE lakes_geo(id serial PRIMARY KEY, the_geog geography(POINT,4326));
```

When you create a table using the geography data type and SRID of 4326, the `geography_columns` view is automatically populated—there is no need to specifically populate it as with the `geometry_columns` table.

If we look at the contents of the view we find our lakes_geo table has been added:

```
geospatial_desktop=# SELECT * FROM geography_columns;
  f_table_catalog   | f_table_schema | f_table_name | f_geography_column | coord_dimension | srid | type
--------------------+----------------+--------------+--------------------+-----------------+------+-------
 geospatial_desktop | hazards        | lakes_geo    | the_geog           |               2 | 4326 | Point
(1 row)
```

You should be aware the PostGIS supports a subset of spatial functions compared to the "regular" geometry type. That's about all we'll say about the geography data type—for more information see the PostGIS manual.

The PostGIS manual is available at http://www.postgis.org

Spatial Index

When working with PostGIS, it's important to make sure you have a spatial index for each layer. Having an index improves both spatial query and rendering performance.

PostGIS provides a Generalized Search Tree (GiST) index for spatial features. Depending on how you created your layers, the index may already exist. You can easily check to see whether an index exists

for a layer using psql to examine the properties of a layer, in this
case the lakes layer we digitized earlier:

```
gsherman@ubuntu:~/geospatial_desktop_data$ psql geospatial_desktop psql (8.4.8)
Type "help" for help.

geospatial_desktop=# \d lakes
                          Table "hazards.lakes"
    Column    |     Type     |                    Modifiers
--------------+--------------+-----------------------------------------------------
 ogc_fid      | integer      | not null default nextval('lakes_ogc_fid_seq'::regclass)
 wkb_geometry | geometry     |
 id           | numeric(10,0)|
 name         | character(80)|
Indexes:
    "lakes_pk" PRIMARY KEY, btree (ogc_fid)
    "lakes_geom_idx" gist (wkb_geometry)
Check constraints:
    "enforce_dims_wkb_geometry" CHECK (st_ndims(wkb_geometry) = 2)
    "enforce_geotype_wkb_geometry" CHECK (geometrytype(wkb_geometry) =
      'POLYGON'::text OR wkb_geometry IS NULL)
    "enforce_srid_wkb_geometry" CHECK (st_srid(wkb_geometry) = 900914)
```

We use the \d command to list the properties of the lakes table.
Notice under the indexes heading there is a primary key on the
ogc_fid field and, under that, a GiST index on the geometry column
wkb_geometry. If you don't see an entry for a GiST index in the list,
you should create one for your table. To create a GiST index for
your table, use the following SQL:

```
CREATE INDEX sidx_lakes on hazards.lakes USING GIST (wkb_geometry GIST_GEOMETRY_OPS);
```

If you import shapefiles using SPIT, a GiST index will not be created. Using shp2pgsql allows you specify the creation of a GiST
index during the import of the shapefile. See Section 13.4, *Importing Shapefiles*, on page 232 for an example. When using ogr2ogr to
import, the spatial index is created automatically.

Loading PostGIS Data

There are a number of ways to load data into a PostGIS-enabled
database. In Chapter 13, *Using Command-Line Tools*, on page 199,
you will see how to load data using both the OGR and PostGIS

command-line utilities. For now we'll look at two other methods for loading data: SQL and QGIS.

Using SQL to Load Data

To load spatial data using SQL, use the GeomFromText function. This function is part of the OGC specification and as an argument takes the Well-Known Text (WKT) representation of a feature. WKT is a simple way to specify a feature type and its coordinates. Some examples of WKT representations are as follows:

- `POINT(-151.5 61.5)`
- `LINESTRING(-151.5 61.5, -151.5 62.5, -152.25 63.0)`
- `POLYGON((-155.82 57.31,-155.94 61.18,-152.82 61.18,-152.78 57.31,-155.82 57.31))`

For example, if we have a lakes layer in WGS 84 (latitude/longitude) coordinates, we would use an insert statement as follows to create a new polygon:

```
INSERT INTO lakes VALUES(1, 'Big Lake', 127.6,
  GeomFromText('POLYGON((-155.82 57.31,-155.94 61.18,
    -152.82 61.18,-152.78 57.31,-155.82 57.31))', 4326));
```

This creates a lake (a pretty rectangular one) named Big Lake with an ID of 1 and a depth of 127.6. The feature type is a polygon, and the spatial reference ID is 4326.

It makes sense to use this method of loading data when you want to import data using a script from a text file or in application code where you are taking input from a user or device. In normal practice, it can become quite cumbersome to manually enter WKT to build up a SQL statement. At least now you are aware of the capability, but you don't need to know this technique to load and use PostGIS data in your desktop GIS application.

Using QGIS to Load Data

QGIS comes with a plugin called SPIT, which stands for Shapefile to PostGIS Import Tool. This plugin allows you to import shapefiles into PostGIS from within QGIS. To use SPIT, make sure it's loaded from the Plugins menu. Once loaded, it will appear as a blue elephant icon on the Plugins toolbar. Before you can use SPIT, you need to have already created a connection to PostGIS (this isn't strictly true, but it makes things easier). We haven't talked about creating connections yet, but you can learn how by jumping ahead to Section 9.4, *Using PostGIS and Quantum GIS*, on page 124. Assuming you have a working connection, just click the SPIT icon to open the tool. In Figure 9.1, you can see the tool with a few shapefiles ready to be loaded into PostGIS.

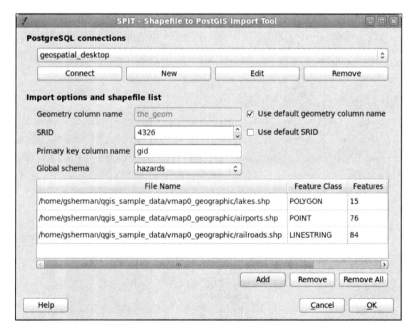

Figure 9.1: Loading shapefiles into PostGIS using SPIT

Let's take a look at each of the items SPIT requires:

PostgreSQL Connections
   The drop-down box lists all the connections you have defined. You can also create a new connection if you don't have one al-

ready. You can also edit the connection selected in the drop-down if things aren't quite right. Connecting to the database will populate the schema drop-down box.

Buttons

There are three buttons to manage the list of shapefiles to be loaded. The list is displayed at the bottom of the dialog box. You can add one or more shapefiles by clicking the Add button and selecting the file(s) from the file dialog box. You can remove a single shapefile from the list by clicking it to select it and then using the Remove button. The Remove All button does as it says and empties the list of shapefiles.

Use Default SRID

Click this checkbox to use the default SRID of -1. This is usually a bad idea because none of your geometries will be associated with a coordinate system as defined in the `spatial_ref_sys` table. It's better to uncheck the box and enter the spatial reference ID in the box.

Use Default Geometry Column Name

If this box is checked, the default geometry column name of "the_geom" will be used when creating tables. If you want to use a different name, uncheck the box, and enter the name you desire.

Global Schema

This drop-down box lists all schemas in your database. If you want to create your new spatial tables in a schema other than "public" (always a good idea), select it from the list.

File List

The file list contains all the shapefiles you have selected for loading. You can edit the feature class and schema for each shapefile entry by clicking the text or choosing the schema from the drop-down list. Editing the feature class type can cause your import to fail but may be needed in some circumstances. The file list also shows the number of features in the shapefile and the name that will be used to create the table.

Once you click the Import button, SPIT proceeds to process each file. A progress bar displays the status as the import proceeds. As the files are processed, they are removed from the list.

Although SPIT is a handy tool, it is also somewhat picky. You may find that, depending on the feature type, some shapefiles can't be loaded. For the fail-safe loading of shapefiles into PostGIS, use one of the methods described in Chapter 13, *Using Command-Line Tools,* on page 199.

Spatial Queries

Let's look at one last feature of spatial databases before we move on to viewing data stored in PostGIS. One of the strengths of an OGC-compliant database is the ability to do spatial queries. PostGIS provides a wealth of both OGC and custom functions to perform queries based on spatial relationships. Using SQL, we can find features that overlap, intersect, touch, or are contained in/by another feature. We can also transform coordinates on the fly, reprojecting them from one spatial reference system to another.

Let's look at one simple example to illustrate. Suppose someone says "I live at latitude 18N, longitude 77W." We want to know where that is—what are our options? We can start up our desktop GIS, load up a world country layer, and move our mouse around to find the location. Or if we have the data in PostGIS, we can quickly do a spatial query to determine the location:

```
# SELECT cntry_name, pop_cntry FROM world_borders
  WHERE GeomFromText('POINT(-77 18)',4326) && wkb_geometry;

 cntry_name | pop_cntry
------------+-----------
 Jamaica    |   2713130
(1 row)
```

The query uses the OGC function GeomFromText to create a temporary point object to use in the search. We use the && operator to test whether the bounding boxes of the features (our point and all polygons in the world) intersect. The query returns the results, in this case Jamaica. This is a simple example of the power of queries using a spatial database. The output isn't a map, and we didn't even use a GUI to answer the question.

For details on using spatial functions and geometry constructors, see the nicely detailed PostGIS manual.[8] Although you may think that using these functions isn't a "desktop GIS" activity, it is an important part of data preparation, conversion, and analysis, so it pays to check out the features and capabilities.

[8] http://postgis.refractions.
net/documentation

9.4 Using PostGIS and Quantum GIS

QGIS and PostGIS have a long history—well, at least from the QGIS side. The first working version of QGIS supported only one data type—PostGIS. So, support for PostGIS has been included in QGIS from day one. This means that the implementation is fairly complete and an important part of the development and maintenance process.

Typical use of PostGIS layers goes something like this:

1. Open the PostGIS dialog box by clicking the Add PostGIS Layer tool.
2. Select the connection to use—if one doesn't already exist, create a PostGIS connection to your database.
3. Connect to the database.
4. Select the layer(s) you want to add to the map.
5. Optionally specify a query to limit the features returned.
6. Optionally set the encoding.
7. Click the Add button to add the layer(s) to the map canvas.

We'll go through these steps one by one. Of course, by this point, we assume you have loaded some data into a PostGIS database. If not, see Section 9.3, *Loading PostGIS Data*, on page 119 for information on loading up your spatial database. To begin, let's open the PostGIS dialog box and create a new connection. Click the New button to open the *Create a New PostGIS connection* dialog box. In Figure 9.2,

you can see the completed dialog box for creating a new connection.
Let's take a look at the required fields.

Figure 9.2: Creating a new PostGIS connection in QGIS

Name

A descriptive name for the connection. This should be unique
enough so you can recall at a glance the database it uses.

Service

The name of a database service defined on the server. This is
setup by your PostgreSQL administrator as a convenience for setting up a connection without having to enter the specifics. We
won't use this in our connection.

Host

The host name where the database resides. This can just be localhost if you are running QGIS on the same machines as Post-
greSQL/PostGIS.

Port

The port number on which the database listens. This is filled in with the default value when you open the dialog box, and you don't need to change it unless your database is listening on a different port.

Database

The name of the PostgreSQL database to which you want to connect.

SSL mode

If supported by the server, the connection can use SSL to encrypt your data so it can't be snooped over the network.

Username

The username used to connect to the database.

Password

The password for the database user.

Save Password

This saves the password along with the rest of the connection information. Depending on your computing environment, this may be a security risk. If you don't store the password, QGIS will prompt you for it at connection time.

Only Look in the geometry_columns Table

Clicking this prevents QGIS from looking through all your tables to see whether they contain a geometry column. This can speed up displaying the list of layers to choose from if you always have an entry in `geometry_columns` for every spatial layer.

Only Look in the 'public' Schema

This constrains QGIS to only look in the public schema when searching for spatially enabled tables.

Also list tables with no geometry

QGIS will include all tables in your database when presenting the list for loading.

Use estimated table metadata

Checking this will cause QGIS to use only a sample of the table and the stored table statistics will be used rather than scanning the table. This is to provide performance when working with large tables—feature count information may not reflect the actual values.

In Figure 9.2, on page 125, we are running QGIS on the same machine as the database, so we specified "localhost." Our database is named "geospatial_desktop," and we are using the standard PostgreSQL port. Once you have filled in the connection information, you can use the Test Connect button to test the connection. If it fails, check the parameters again. If they are correct, you may have to check the PostgreSQL database access configuration to make sure you have privileges to connect. Once you can connect, just click OK to save the connection. This takes you back to the *Add PostGIS Table(s)* dialog box.

We are now ready to connect to the database—with our new connection selected in the drop-down list, just click the Connect button. Once you do this, the list of available layers is populated as shown in Figure 9.3.

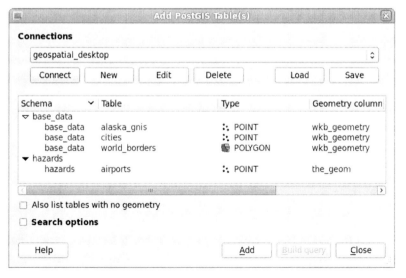

Figure 9.3: List of available PostGIS layers

If we look at the list of tables, we see a representative collection of

PostGIS data layers in our database, organized by schema. Under the Type column, you will notice an icon that indicates the feature type stored in the table. These can be point, multipoint, linestring, multilinestring, polygon, or multipolygon. You can't distinguish from the icon whether a given feature is a regular or "multi" type feature.

> **What's the Difference Between a Table and a Layer?**
> ───
> In our discussion, a PostGIS layer is a table in PostgreSQL that has a geometry column. It may or may not have a record describing it in the geometry_columns table.
> So, layers are also tables, and you may find us referring to them in both ways. A PostgreSQL table without a geometry column is just that—a table in the database.

For example, the world_borders layer is a polygon layer in the "base_data" schema and has its geometry stored in a column named wkb_geometry. If you scroll the layer list you'll notice there is a "Sql" column on the right. This will be initially be empty, and we'll see its purpose in just a minute.

Loading a layer from the list is easy. Just select one or more by clicking them (they will be highlighted as you click so you know they are selected); then click the Add button. The layers will be added to the QGIS map canvas and drawn using a random color. Once loaded, you can modify the colors and rendering using the symbology options we discussed in Section 5.4, *Advanced Viewing and Rendering*, on page 48.

Suppose we have a PostGIS layer with 10 million features. As you can imagine, it would take a while to draw, moving all the data across the network. Or consider a layer that contains thousands of features, but we are interested only in some of them based on one of their attributes. This is where the ability to limit the features in a PostGIS layer comes in handy. You could think of these as "virtual layers" since they are defined by a query at the time you add them to

QGIS. Let's look at an example using the Geographic Names Information System (GNIS) available from the U.S. Geological Survey.[9]   [9] http://geonames.usgs.gov

GNIS contains information about geographic features, including the "official" name. For example, the data includes lakes, streams, islands, glaciers, towns, and schools. All the features are represented by a single point. We'll use the GNIS data for Alaska in our example and add several "virtual layers" based on queries against the alaska_gnis table.

First we open the *Add PostGIS Table(s)* dialog box, connect to our database, and scroll through the list of layers until we find the alaska_gnis layer. Instead of clicking it and adding it to the map, we double-click to open the PostgreSQL Query Builder. You'll notice it's pretty much the same as the query builder we used in Section 5.5, *Advanced Search*, on page 65. The difference of course is that now we are querying a real database instead of a shapefile. In Figure 9.4, on the following page, you can see the query builder populated with the parameters for our first layer (schools) and the results of clicking the Test button.

The query we did to create schools returned 103 rows. Once we click OK in the query builder, we are returned to the *Add PostGIS Table(s)* dialog box. Note that now there is something in the "Sql" column next to the alaska_gnis layer. This is just the contents of the query box but serves to remind us what we are adding in the event that we set up queries and add more than one layer at a time. With the alaska_gnis layer selected, we click the Add button to add it to the map. In Figure 9.5, on page 131, you can see that in addition to the school layer, we added layers for airports and mines. QGIS doesn't provide a very pleasing name in the legend when adding layers in this way, so we took the liberty of renaming each of the GNIS layers to something sane. So, now we have a map with three separate layers, all derived from the alaska_gnis layer in our database.

Now maybe you are asking yourself, why not just add the alaska_gnis layer and symbolize it based on type? We could do that, and it

Figure 9.4: PostGIS query builder in QGIS

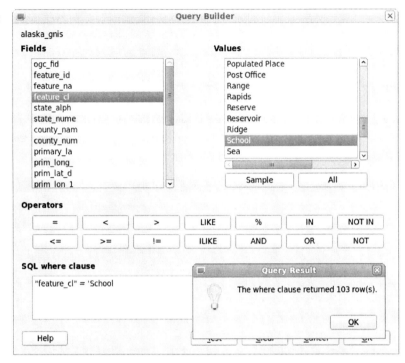

might work assuming the following:

- Our data is not too dense.
- We want to see all types, not just schools, airports, and mines.
- Our layer isn't so large that it causes performance problems.

In the case of the GNIS data, symbolizing all of it by type would result in a blob of dots and colors, with lots of overlap. We could remove classes so only the three we are interested in show up, however, there are 63 types in the database and this would be a bit of work. Using the same PostGIS layer to create our "virtual" layers turns out to be an efficient way to get just the data we want out of a large dataset. Though we didn't demonstrate it, your queries to create a layer can be more complex than just a simple `this='that'` query by using operators such as AND and OR to select rows on more than one condition.

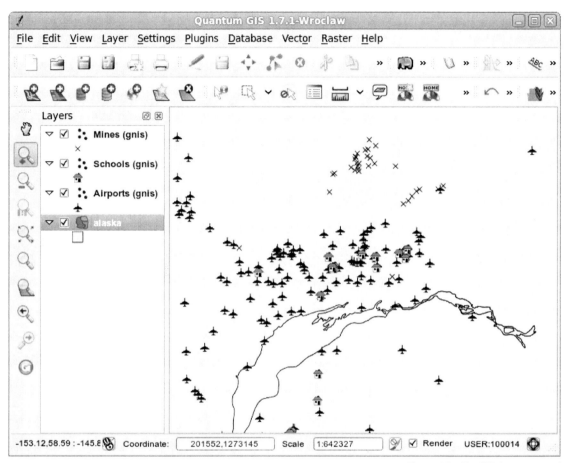

Figure 9.5: PostGIS layers created with the query builder

Before we leave this topic, we should mention one more thing. Once you have created a layer using a PostGIS query, you can change it using the Query Builder button found on the General tab of the vector Layer Properties dialog box. Clicking the button opens the query builder, allowing you to modify (or completely change) the query that defines the layer.

Creating a Spatial View

If you are a SQL wizard (or wizard-in-training), you can accomplish the same effect as our "virtual layers" using database views.

Whether you choose to do this depends on how frequently you need
to access the filtered data. If you always use a certain subset of a
given layer, creating a spatial view is a good solution. For example,
to make our schools layer always available, we can create a view using psql, the PostgreSQL interactive terminal (of course, you could
use any tool that can access PostgreSQL and execute queries).

```
gsherman@ubuntu:~/geospatial_desktop_data$ psql geospatial_desktop
psql (8.4.8)
Type "help" for help.

geospatial_desktop=# CREATE view school_view as SELECT * FROM alaska_gnis WHERE feature_cl = 'School';
CREATE VIEW
geospatial_desktop=#
```

This creates a view for us that includes all the columns from the
alaska_gnis table but includes only those features that are schools.
When you fire up QGIS and connect to the database, you'll find the
school_view in the list of available PostGIS layers.

9.5   Using PostGIS and uDig

You can use uDig to display PostGIS layers—which is no surprise
since both come out of Refractions Research.[10] If you look back
to Figure 5.1, on page 42, you'll recall that PostGIS was one of the
choices when adding data to the map.

Adding a PostGIS layer is pretty easy—you just have to know the
particulars of your database location and connection parameters,
just as we did with QGIS. uDig provides a step-by-step process to
connect to a PostGIS database and load a layer. Once the connection
is made and we click Next, we are presented with a list of layers in
the database that can be added to the map. uDig doesn't currently
support the filtering of PostGIS layers, so we can't create a "virtual"
layer. Once the layer is added to the map, you can symbolize it
just like we discussed in Section 5.2, *Rendering a Story*, on page 44,
including the use of color palettes.

Figure 9.6, on the facing page shows the cities layer loaded over
the world_borders.

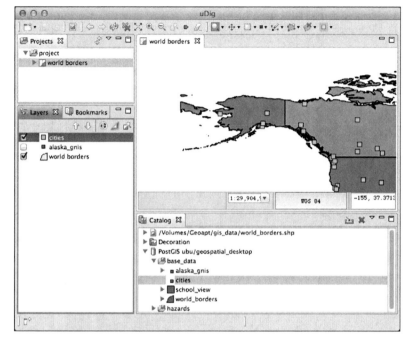

Figure 9.6: PostGIS connect dialog box in uDig

Once you've made a connection, uDig keeps it available in the cat-alog, accessible at the bottom of the workspace. When you click the Catalog tab, you'll get a list of the data stores available to you, one of which will be your PostGIS connection(s). If you expand the PostGIS node, you'll see a list of all the layers for a given connec-tion. To add one of the layers, simply right-click it and choose Add to New Map or Add to Current Map. You'll note this is also a quick way to create a new map and get some data on it. If you choose Add to New Map, a new map tab is created and named the same name as the layer you chose.

9.6   SpatiaLite

We looked at server based spatial databases—now let's take a very brief look at a serverless solution. SpatiaLite is a self-contained, file-based spatial database extension for SQLite. The website declares it "a complete Spatial DBMS in a nutshell".

SpatiaLite allows you to store all your data related to a single project in one database file. This makes it highly portable. It provides support for OpenGIS functions for spatial analysis and query. This allows you to examine the spatial relationships between features. It supports both WKT and WKB geometry formats. Spatial indexes are implemented using SQLite's RTree extension.

To import shapefiles, use the spatialite-gui application.

Two "virtual" extensions (VirtualShape and VirtualText) allow you to access and query shapefiles and CSV files as SQLite virtual tables without having to import the data into SpatiaLite. Shapefiles can also be imported and exported with SpatiaLite. Additional formats can be imported using ogr2ogr.

QGIS provides full support for SpatiaLite out of the box. The user interface allows you to create a new SpatiaLite layer and even create a brand new database in the process. This functionality can be accessed from the Layer→New SpatiaLite Layer... menu. SpatiaLite layers are loaded using the Add SpatiaLite Layer tool on the Manage Layers toolbar (the same one used to load vector, raster, and PostGIS layers). Once loaded, a SpatiaLite layer behaves like any other vector layer in QGIS.

[11] http://www.gaia-gis.it/spatialite

SpatiaLite is great for bundling your data and having a full-featured spatial database that works across Linux, Mac OS X, and Windows. For more information see the SpatiaLite website.[11]

9.7 Summing It Up

You now have been exposed to the power and flexibility of a spatially enabled database. Should you use a spatial database or stick to file-based data like shapefiles? That depends on your needs and goals. If you have large datasets that you want to create "virtual" layers from using views or definition queries, a spatial database is the way to go. Another good reason is to create a centrally located, shared data source for multiple users.

A spatial database adds a bit of complexity in terms of getting started, but it's worth the effort when managing large datasets and many layers. If you are a casual user, you may find it's not for you—again, it depends on your goals and needs.

Lastly, you may be wondering why we are talking about server software in a desktop book. If you've gotten this far, you realize that the "back end" is just as important as the front. Using a spatial database provides a data store that we can use on the desktop, as well as for web mapping applications. From that perspective, it's a good choice as a central repository for all our data.

10

Creating Data

Using existing data is fine and gives us a lot of capability—until we want to display data specific to our area of interest. Sometimes we luck out and find the data; other times we have to create or convert it. At some point in your OSGIS career, you are going to need to do some creation or conversion of data to get what you need. This is where you move on from the hunter-gatherer stage in your GIS data usage.

Ways to create data suitable for our use include the following:

- Digitizing
- Importing from text files or other sources
- Converting data
- Importing GPS data
- Georeferencing an image

In this chapter, we'll explore some of the ways in which we can torture data (whether raw or cooked) into submission and make it usable.

10.1 Digitizing

We've already seen examples of digitizing in the previous chapters. In its non-glamorous form, digitizing is just tracing features, whether it be on a digitizer tablet or the screen. This is a tried-and-true method of generating new vector data from paper or a scanned image. If you digitize from a scanned image displayed on your screen (as we did in Chapter 7, *Digitizing and Editing Vector Data*, on page 89), it's called *heads-up digitizing*—you've got to keep your head up and focused on the monitor to do it. This method of digitizing has become quite prevalent with the availability of imagery and the ability to scan large documents and maps into a format suitable for onscreen display.

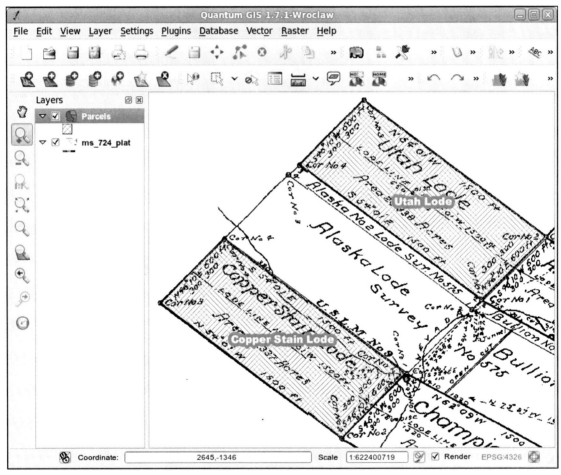

Figure 10.1: Digitizing a plat

As an example, we went out on the Internet and dug up a 1906 mineral survey plat from Alaska. Our task is to create a vector layer from the plat for use in historical archiving or some other creative purpose. Of course, digitizing plats is an ongoing activity for government entities. Our plat is a TIFF image and is not georeferenced. Since we don't have any reference points to register it, we'll pretend it's in the proper coordinate system. If we were doing this for real and wanted the vector layer to overlay other physical features, we would need to get it georeferenced first.

In Figure 10.1, on the facing page, you can see our work partially complete, with the completed parcels crosshatched and labeled. The plat itself is a black-and-white scan of an original paper plat. We created a new shapefile using QGIS and began digitizing the plat, storing an id and the claim name for each parcel in the attribute table as we go. When complete, this gives us a new vector layer that contains the id and name for each parcel. This in turn can be displayed with other vector layers in the same coordinate system or linked by parcel ID to additional data in a database.

An alternative to digitizing the raster is to use the GRASS r.to.vect command. This will create a vector layer from the raster. The results depend on the quality of the raster. In the case of our plat, we would end up with a huge vector layer containing tens of thousands of polygons. When using this method, everything on the image gets converted, including the text. On our example plat, the quality of the lines around the individual lots is such that we end up with an inner and outer polygon, the outer one being as wide as the line on the image. In addition, the process creates a polygon for the entire image boundary. You could spend as much time cleaning up the result as digitizing the polygons from scratch.

To aid in cleaning up the vectors, you can use v.clean with the tool=rmarea tool to remove small areas. The other option may be to pre-process the image using a graphics program to remove some of the noise and unwanted information. In any case, you may find that r.to.vect is an effective solution when you need to vectorize a raster.

Digitizing is an activity that you'll find insanely boring, tedious, interesting, or therapeutic, depending on your outlook. It remains an important means to create vector data from raster.

10.2   Importing Data

Another important way to get data into your GIS realm is by importing it from text or other source files. Depending on the format of the data, you may find there is a ready-made solution for importing it. A prime example of this is delimited text that can be easily imported by both QGIS and GRASS. QGIS supports the import of points only, while GRASS can accommodate all feature types in the GRASS vector model.

Quite often you find yourself with some text data that contains coordinates and other attribute information that's begging to go into your GIS. Let's start with a simple example to get started. We want to create a data layer of all the volcanoes in the world. Using a search engine, we find a website[1] that provides a means to search for volcanoes and their locations. By not entering any search parameters, we are presented with a web page containing a table of all volcanoes and their locations.

[1] http://www.ngdc.noaa.gov/hazard/vol_srch.shtml

Since the website doesn't provide a download of the data, the first thing we need to do is copy and paste the results into a text editor. Doing this provides us with a text file containing a header row with the field names, followed by a row for each volcano:

```
Number  Volcano Name  Region  Latitude  Longitude Elev  Type  Status  Last Known
Eruption
0803-001 Abu Japan Honshu-Japan 34.5 131.6 571 Shield volcano Holocene Unknown
1505-096 Acamarachi Chile Chile-N -23.3 -67.62 6046 Stratovolcano Holocene Unknown
1402-08= Acatenango Guatemala Guatemala 14.501 -90.876 3976 Stratovolcano Historical D1
0103-004 Acigol-Nevsehir Turkey Turkey 38.57 34.52 1689 Maar Holocene U
1505-017 Acotango Bolivia Chile-N -18.37 -69.05 6052 Stratovolcano Holocene U
```

The file looks a bit scrambled up with no clear spacing or delimiter. Looking at the text file in our editor, we discover that the columns of the table are separated by a tab character. We can use tab as our delimiter to import the data. The only change we need to make is to clean up the header row (the first line of the file). We can modify the field names to shorten them and make them more appropriate for import. The other change is to delete the second line of the file, since the Last Known Eruption field name is broken across two lines. Our changed header now looks like this:

```
Number Name Country Region Latitude Longitude Elev Type Status Last_Eruption
```

When making the changes, make sure that each field name is separated by a tab character. Otherwise, the import won't work properly.

Importing Data with QGIS

With the header row fixed, we are ready to import the data. First we will use the QGIS Delimited Text plugin to load and view the data. From the QGIS Plugin Manager, load the plugin to add the tool to the Plugins toolbar. Click the Add Delimited Text Layer tool to begin the import.

For an overview of plugins in QGIS, see Section 19.4, *Plugins*, on page 372.

In Figure 10.2, on the following page, you can see the delimited text plugin dialog box, populated with the parameters for our input file.

We used the browse button to populate the delimited text file box with our prepared file. The plugin is actually pretty smart. When you enter the filename, the remainder of the parameters for the import are populated using an educated guess. When you set the delimiter, in our case tab, the plugin parses the data and fills in the X and Y fields if it can. The sample text box shows you what's happening so you can adjust the delimiter(s) accordingly. For our volcanoes file all we have to do is click Tab as the selected delimiter and clear the other checkboxes. Notice that Latitude and Longitude were filled in automatically. If not, you can select the correct fields from the drop-down boxes. To add the layer to the map, click OK.

Once you've added the text file to the map, you can use it just like any other layer, including identifying features and viewing the attributes. So far we haven't really imported anything. The Delimited Text plugin includes a *data provider* that allows QGIS to treat the text

Figure 10.2: The QGIS Delimited Text plugin

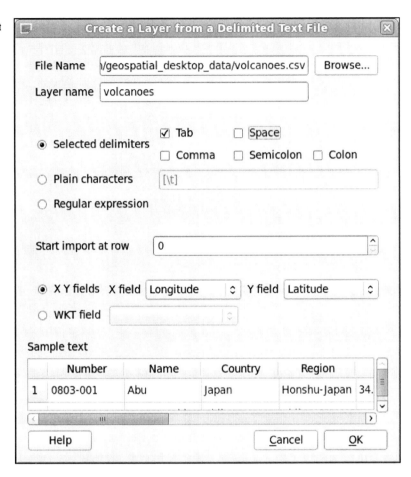

file like a true layer. Essentially a data provider acts as a translator between QGIS and the data store, whether it be a text file, OGR layer, or PostGIS layer. If you are happy with the text file, you can save it as a shapefile by right-clicking the layer in the legend and choosing Save as.... The next time you want to use the data, just load the shapefile rather than going through the text import process.

Keep in mind that you can use anything for a delimiter—it doesn't have to be a tab character. QGIS also supports the use of regular expressions when defining the delimiter, allowing you to import text data that is not entirely uniform.

Preprocessing Text for Import

Sometimes the data you find may not be ready for import. When this happens, you're faced with preprocessing; either in an editor or with a script. Let's look at an example that illustrates this point.

Say we want to plot historic earthquakes in Alaska. To get started, we can download earthquake data for Alaska[2] from 1898 through 2006. Upon the examination of the data, we find that it is a fixed-length format. Some columns have blanks in some rows and values in others. This means we can't just split the record up on the whitespace to get the fields.

[2] http://www.aeic.alaska.edu

```
Mo/Dy/Year  Hr:Mn:Sec    Latitude   Longitude   Depth    mb     ML     MS
                                                 (km)
12/23/1906  17:21:11.7   56.8500 N  153.9000 W   0.0            7.3    7.3
08/22/1907  22:24:00.0   57.0000 N  161.0000 W  120.0    6.5    6.5
05/15/1908  08:31:36.0   59.0000 N  141.0000 W   25.0            7.0
```

Note that the "mb" column has a value in the second row but not the first and third.

To properly parse the records and get the fields we want, we resort to writing a small Ruby script to prep the data. In addition to breaking it out by fields, we also check for longitudes in the western hemisphere and set them to negative to make sure they plot where they should in the world.

parse_earthquakes.rb

```ruby
#!/usr/local/bin/ruby
# Prep a text file of earthquake events with fixed length records to be
# imported as delimited text. The "|" is used as the delimiter.
#
f = File.open("../db_search10423")
# Skip the first two header records
2.times {f.gets}
# print the delimited header row containing the fields we are interested in
print "event_date|event_time|latitude|longitude|depth|magnitude\n"
# process the earthquake records
while not f.eof
  record = f.gets
  # use a fixed length approach to get the fields we want since
  # splitting on white space isn't feasible
  event_date = record[1..10]
```

```ruby
    event_time = record[13..22]
    latitude = record[26..32]
    longitude = record[37..44]
    longitude_direction = record[46..46]
    depth = record[50..54]
    magnitude = record[66..69]
    # if the longitude is in the western hemisphere, it must be
    # negative
    if longitude_direction == 'W'
      longitude = -1 * longitude.to_f
    end
  end
  # print a delimited record
  STDOUT << event_date << "|" << event_time << "|" \
    << latitude << "|" << longitude << "|" \
    << depth.strip << "|" << magnitude.strip << "|\n"
end
# close the input file
f.close
```

When we run this script, we get a nicely formatted file, delimited with | and containing only the fields in which we are interested.

```
event_date|event_time|latitude|longitude|depth|magnitude
06/29/1898|18:36:00.0|52.0000|172.0000|0.0|7.6
10/11/1898|16:37:32.7|50.7100|-179.5|0.0|6.9
07/14/1899|13:32:00.0|60.0000|-150.0|0.0|7.2
```

You can now import the data using the Delimited Text plugin in QGIS, using the same method as we used with the volcano data. The only difference is this time we are using | as a delimiter.

If you are lucky, you won't have to go through a big preparation process before importing your data. Oftentimes you can patch up the text file using a text editor and global search/replace to get it formatted for import. When you can't just write a quick Ruby (or Python or Perl) script to do the job.

Importing with GRASS

Once we have the delimited text file formatted, importing into GRASS is even quicker than using QGIS. From the GRASS shell, we can use the v.in.ascii command to import the data. If you don't specify column names, GRASS assigns default names that may not be too meaningful. We want meaningful names, so for the columns option, we specify the name for each in SQL style.

```
> v.in.ascii input=earthquakes_delim.txt output=earthquakes skip=2 \
x=4 y=3 cat=0 columns="event_date date, event_time varchar(10), \
lat double precision, lon double precision, depth double precision, \
magnitude double precision"

Scanning input for column types...
Maximum input row length: 57
Maximum number of columns: 6
Minimum number of columns: 6
Column: 1 type: string length: 10
Column: 2 type: string length: 10
Column: 3 type: double
Column: 4 type: double
Column: 5 type: double
Column: 6 type: double
Importing points...
 100%
 100%
Populating table...
Building topology for vector map <earthquakes>...
Registering primitives...
14377 primitives registered
14377 vertices registered
Building areas...
 100%
0 areas built
0 isles built
Attaching islands...
Attaching centroids...
 100%
Number of nodes: 14280
Number of primitives: 14377
Number of points: 14377
Number of lines: 0
Number of boundaries: 0
Number of centroids: 0
Number of areas: 0
Number of isles: 0
v.in.ascii complete.
```

The options to the v.in.ascii command are explained in the GRASS
manual, but basically apart from the input and output names, we
told the command to skip the first two lines since they are header
lines and that our x coordinate is in column 4 and the y coordinate
is in column 3 of the input file. We also specified cat=0 to indicate we wanted GRASS to create an ID column for us. If the input

file had a suitable ID field, we would have used it by specifying its column number with the `cat` option.

Notice we didn't specify the | delimiter when using `v.in.ascii`. That's because it is the default delimiter. If we had used a different delimiter when preparing the text file, we would need to use the `fs` parameter to specify it.

Once we have our new earthquakes layer imported into GRASS, we can symbolize it by magnitude to see where in Alaska we shouldn't live. In Figure 10.3, on the next page, you can see a portion of south-central Alaska with the earthquakes symbolized by magnitude.

As you can see, it's relatively easy to import text data into QGIS or GRASS. Our examples dealt only with importing points. As I said earlier, GRASS supports importing all feature types in "standard" mode. See the manual page (`g.manual v.in.ascii` from the command line) for details and examples on importing types other than points.

10.3   Converting Data

Sometimes we are faced with converting data before we can use it, simply because it's not in the format that works with our software. You can probably think of other reasons as well; for example, you want to share the data with someone using another type of software. Or perhaps you might want to transform the data to another projection to make it play nicely with your other data. You might recall we talked a bit about standardizing your data format in Chapter 8, *Data Formats*, on page 103.

A lot of times data is distributed in a form that's convenient for the distributor, not the end user. This is probably the most common circumstance you'll encounter when gathering data from the Internet. Let's take an example.

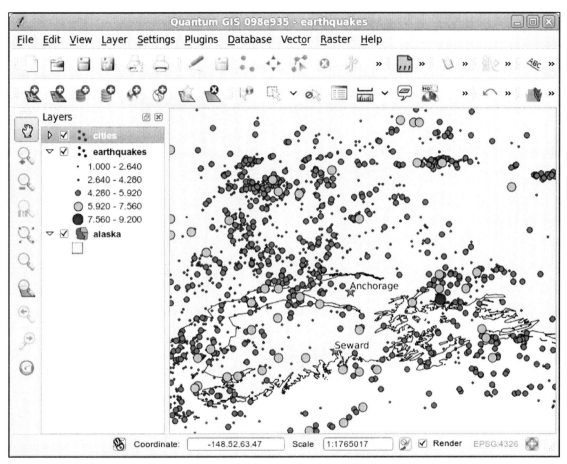

Figure 10.3: Earthquakes rendered in
QGIS by magnitude

Importing an E00 Interchange File

Harrison is interested in some data from his local state government.
He soon discovers that they, like a lot of other government entities,
deliver their data in a format called an E00 file.[3] To make use of the
data, Harrison has to convert it into a format his software supports.
He has several options for converting the interchange file. First he
can search around the Internet for one of several E00 to shapefile
converters. The other option is to use GRASS to import it right
into his favorite mapset.[4] Once in GRASS, he can also export it in

[3] This is an ArcInfo interchange format
that is not directly usable by most GIS
software.
If you need help getting GRASS
started, see Appendix 18, on
page 323.

[4] The GRASS conversion requires
avcimport and e00conv utilities—see
http://avce00.maptools.org

a number of formats to share with his friends. To convert an E00
interchange file to GRASS, Harrison uses the following:

```
GRASS 6.4.2RC1 (alaska_albers):~> v.in.e00 file=beikman.e00 type=area \
  vect=beikman_map
An error may appear next which will be ignored...
E00 Compressed ASCII found. Will uncompress first...
...converted to Arc Coverage in current directory
Importing areas...
Over-riding projection check
Layer: LAB
WARNING: Column name changed: 'BEIKMAN#' -> 'BEIKMAN_'
WARNING: Column name changed: 'BEIKMAN-ID' -> 'BEIKMAN_ID'
Counting polygons for 5069 features...
Importing map 5069 features...
 100%
Layer: ARC
WARNING: Column name changed: 'BEIKMAN#' -> 'BEIKMAN_'
WARNING: Column name changed: 'BEIKMAN-ID' -> 'BEIKMAN_ID'
Counting polygons for 11131 features...
Importing map 11131 features...
 100%
--------------------------------------------------
Building topology for vector map <beikman_map_tmp>...

... more messages ...
Imported <area> vector map <beikman_map>.
Done.
GRASS 6.4.2RC1 (alaska_albers):~ >
```

If Harrison needs to share the new layer with somebody, he quickly
creates a shapefile using `v.out.ogr`:

```
GRASS 6.4.2RC1 (alaska_albers):~ > v.out.ogr -c input=biekman_map type=area dsn=. \
  olayer=beikman_map format=ESRI_Shapefile
Exporting 5069 areas (may take some time)...
 100%
v.out.ogr complete. 5069 features written to <beikman_map>
(ESRI_Shapefile).
GRASS 6.4.2RC1 (alaska_albers):~ >
```

You can also convert an E00 to
an ArcInfo binary coverage using
`avcimport`.   The result can be
opened directly with QGIS.

You'll find that GRASS supports a lot of input and output conversions for both getting new data into GRASS and exporting it out for
use with other applications or to share with others.

Another great way to convert both vector and raster formats (including e00) is using the GDAL/OGR suite of tools. Since there are

so many possibilities, we'll explore these tools in Section 13.2, *Using GDAL and OGR*, on page 212.

10.4   Using GPS Data with QGIS

GPS units are everywhere these days. Between the practical use for the professional and the recreational user, as well as the popularity of geocaching,[5] it seems like everybody is using them. I'm sure you would like to display your GPS adventures on a topographic (read DRG) map, especially after we work through how to create seamless rasters using GRASS in Chapter 12, *Geoprocessing*, on page 171. Well, the good news is there are lots of open source tools available for working with your GPS. You'll need a GPS with an interface cable so you can move data to and from your computer. In this section, we'll show you how to put your data on the map using the GPS plugin that comes with QGIS.

[5] Treasure hunting with a GPS. See http://www.geocaching.com.

Getting Set Up

Obviously you need QGIS installed and working on your platform. The only other requirement (apart from a GPS unit) is gpsbabel.[6] QGIS uses gpsbabel to import other formats and for GPS downloading and uploading operations. Fortunately, gpsbabel runs on all the same platforms as QGIS, so if you can run QGIS, you can use it with your GPS. Actually, gpsbabel is a remarkable little program. It runs on almost everything and supports 100+ formats and a bunch of GPS hardware. For upload/download, if you have a Garmin or Magellan unit, you should be good to go. For others, see the format list on the web page to see whether yours is listed there.

[6] http://www.gpsbabel.org

The GPS Plugin

The GPS plugin is part of the core distribution of QGIS. This means it is included by default with your installation of QGIS. Like other plugins, you have to load it before you can get at it from the Plugins toolbar. If you don't know how to load a plugin in QGIS, take a look at Section 19.4, *Plugins*, on page 372. Once you have the plugin loaded, you'll see a nice little GPS icon on the Manage Layers toolbar. The plugin also adds a couple of entries to the Plugins→GPS menu. Clicking on the GPS tool or choosing Plugins→GPS→GPS Tools from the menu will open the interface you see in Figure 10.4.

The first thing to notice is the five tabs across the top of the GPS plugin dialog box. The first two (Load GPX File and Import Other File) allow you to load data stored on a disk, a CD, or maybe a thumb drive. The next two tabs (Download from GPS and Upload to GPS) provide the tools needed to move data to and from your unit. Lastly we have the GPX Conversions tab which allows you to perform conversions from a gpx file. Now that we have the plugin loaded up and ready to go, let's begin by fetching some GPS data from our unit so we can display it.

What is GPX?

GPX stands for GPS Exchange Format. It's a lightweight XML format for interchanging your GPS data between applications, both on the desktop and the Web. GPX can handle waypoints, tracks, and routes. Using the GPX format, you can easily exchange your data between a host of GPS and GIS applications. QGIS directly supports the GPX format using the GPS data provider that is included with all versions of QGIS. For a list of applications that can work with the GPX format, take a look at the TopoGrafix website at www.topografix.com/gpx_resources.asp.

Downloading Data from Your GPS

OK, so we assume you have your GPS hooked up to the computer with your interface cable and the unit is powered on. For now we'll also assume that it's a Garmin, because the plugin is set up for that already. Downloading from your GPS is easy.   Just click the Download from GPS tab, and select the port your interface cable is using. The port names will vary depending on your operating system. Usually you'll find the appropriate port in the drop-down list.

For help on port names, see the QGIS Manual.

Since QGIS is feature oriented, you have to specify whether you are downloading waypoints, routes, or tracks. Just choose what you want from the drop-down box.

When you download from your GPS, QGIS creates a new layer and also saves the data in a new GPX file. For it to do that, you have to provide an output filename and a name for the layer.

When you click OK, the data is pulled from the GPS unit and stored on disk, and a new layer is added to the QGIS map canvas. Repeat the process to fetch each of the feature types you want to download (waypoints, tracks, and routes). In Figure 10.5, on the following page, you can see our GPS data added to the map. The track is displayed as a wide, dashed line and the waypoint as a simple circle marker. To make it more interesting, we've added a Google map hybrid layer using the OpenLayers plugin. We've also labeled our lonely little waypoint using the data that came with it from the GPS, in this case, MOOSE. Once loaded into QGIS, the GPX data behaves just like any other layer. You can label it, change the symbol type and colors, and rename it in the legend. In case you're wondering, we called the waypoint MOOSE because that location looked like really good moose habitat.

The OpenLayers plugin is a Python plugin that supports Google, Open-StreetMqp, Yahoo, and Bing layers.

Loading and Viewing Data

So far, we have pulled the data from our GPS unit, displayed it, and in the process created a GPX file on disk. The next time we want to

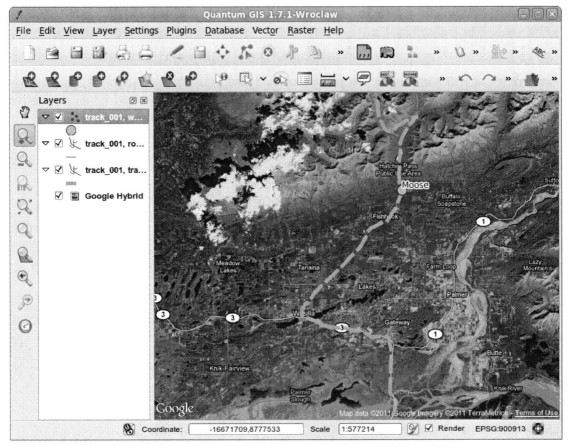

Figure 10.5:   Track and waypoint loaded from GPS unit

display it, we don't have to pull it off the GPS again (well we could, but only if we felt we needed the practice). We already have the file saved to disk—to load it we have two options: use the GPS plugin or the regular Add Vector Layer tool.

From the plugin, we just click the Load GPX File tab if it's not already active and then enter or browse to the location of our gpx file. Using the three checkboxes below the filename, we can choose to load all the feature types from a GPX file or just some. For each box checked, you will get a separate layer in QGIS.

You may have noticed that the GPX files we created by downloading from our GPS contain only one feature type each. You likely also noticed the three checkboxes when loading a GPX file. QGIS allows you to load multiple feature types because the plugin assumes you may have GPX files from other sources that contain multiple types.

If we use the Add Vector Layer tool to load our GPX file we first select the file (change the file filter in the *Open an OGR Supported Vector Layer* dialog box to GPX) and then choose which feature types from the list that pops up. The list shows you what's available in the GPX file as well as how many features in each. Once you select the feature type(s), click and click OK the layer(s) are loaded. Which method you use to load GPX layers is strictly a matter of preference.

The GPS plugin also allows you to load other formats supported by gpsbabel. Say you have a batch of files from a GPS that you want to view. If you click the *Import Other File* tab, you can convert them to GPX format so they can be used with QGIS. The real work here is done by gpsbabel, so you must have it installed on your system in order for this to work. This is just a handy way to convert files and get them into QGIS. Of course, you could also just use gpsbabel from the command line to accomplish the same task.

Uploading Data to the GPS

The last thing we'll look at is uploading data from QGIS to your GPS unit. Here are a couple of reasons why you might want to do that:

- You download some routes for trails in your area from your local parks department, and you want to load them on your GPS.
- You have edited your waypoints and tracks from your GPS and want to load them up.

In the second scenario, you'll notice that yes—you can edit your GPX layers in QGIS and change not only the location (more on that in a second) but also the attributes. So if you're like me and have trouble entering text for waypoints on the little spongy rubber buttons, you can now edit the things after the fact to correct and enhance them. You may be thinking that editing the locations kind of violates the intent of a GPS. Well, there may be some circumstances where you might want to move a waypoint, for example, if you have better information gleaned from a data source with better accuracy than your GPS.

The other thing you can do is add new features. So, for example, you could digitize trails from a DRG and upload them to the GPS. Of course, you need to create a new GPX file using the Plugins→GPS→Create new GPX layer menu. Once you're done, you can hit the trails confident that you won't get lost (depending on how good your digitizing was). Naturally, the better way is to download the data from the local parks department and put it on your GPS, but sometimes that information isn't available, especially in a GPS format.

For this to work you must add the GPX layer(s) to the map using the plugin, not the Add Vector Layer tool

If we have some data we want to upload, we can just open the plugin and click the *Upload to GPS* tab. From the Data Layer dropdown list, you can select the GPX layer you want to upload. Notice that it has to be a layer that's already on the map. You can't upload an arbitrary file using the plugin. Once you select the device and port, just click the OK button and watch the data fly onto your GPS. In Figure 10.6, you can see the plugin ready to upload a routes layer to our GPS unit.

Figure 10.6: Uploading to the GPS

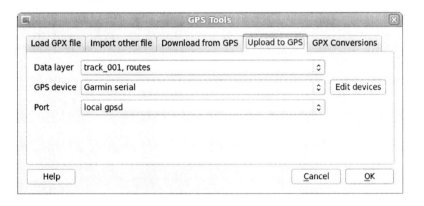

Remember, the GPS plugin relies on gpsbabel to do the heavy lifting so make sure you have it installed on your system before working

with your GPS.

10.5   *Georeferencing an Image*

In case you haven't picked up on it by now, a georeferenced image is one that has associated coordinate information such that it "draws" where it belongs in the world. An image that isn't georeferenced in GIS is not much better than a photograph. You can still look at it, but you can't really do anything GIS-like with it, such as overlay vector data or digitize from it.

Georeferencing with QGIS

QGIS includes a plugin to georeference an image, provided you have the X and Y coordinates needed to establish control points. A control point is a point on the image that you can determine the real-world coordinates for with a high degree of accuracy. The more accurate your control points, the better the "fit" of the georeferenced image.

To begin, load the Georeferencer plugin from the QGIS Plugin Manager. To start the process, click the Georeferencer tool on the plugin toolbar. The Georeferencer has its own map canvas, along with map navigation tools. Load the raster (image) file you want to georeference using the Open raster tool on the toolbar. Once the raster is displayed, you can add control points. Zoom in on the image to the location of one of your control points, click the Add Point button in the toolbar, and then click the map to add the point. When you click, you will be prompted for the X and Y coordinates for the control point. If you are fortunate to have a good control layer loaded in QGIS, click the *From map canvas* button and click on the QGIS canvas to set the coordinates.

Continue to add control points, preferably covering the extents of the raster. In Figure 10.7, on page 157, we have loaded an image and added four control points by manually entering the X and Y coordinates.

Once the control points are established, choose Settings→Transformation settings from the menu or just click the tool on the toolbar. Figure 10.8, on page 158 shows the settings dialog for our transformation. In this case we just specified the transformation, sampling method, output raster, and the projection.

The Georeferencer allows you to save and later load your control points. This is a good idea in case you need to tweak the results later. Depending on the transformation type you choose you can use the Generate GDAL script tool on the toolbar to create a command line script you can run to do the transformation. You might want to save the script for future reference. To do the actual transformation from the Georeferencer, click the Start georeferencing button. Once the transformation is complete, load the output raster in QGIS and check it against any reference layers you may have to make sure the process was a success.

Georeferencing with GRASS

There are three ways to georeference an image using GRASS.

- Command line using i.group + i.target + i.points/i.vpoints + i.rectify
- The tool in the Tcl/Tk gis.m GUI file menu
- The wxGUI Georeferencer

[7] http://grass.osgeo.org/wiki/Georeferencing

For information on each of these methods, see the GRASS Wiki[7]

The process is pretty much like that we used with QGIS. We won't look at an example here, but you can explore the command line and GUI methods on your own if you decide to use GRASS for georeferencing rasters.

Of course, the best option is to find rasters that are already georeferenced. In many cases, you can do this—if not, you have to use your newly acquired skills to get the job done.

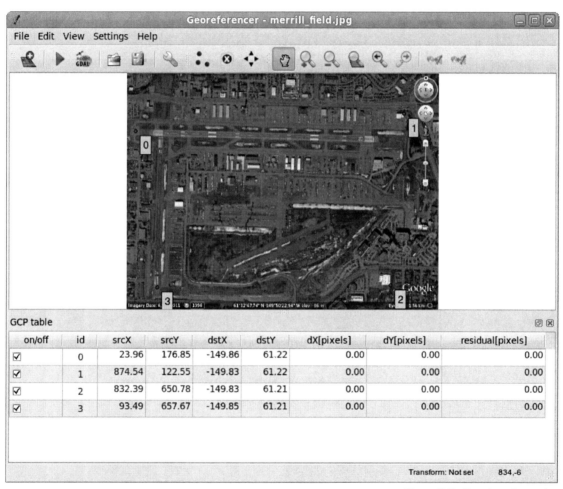

Figure 10.7: Georeferencing an image with QGIS

Figure 10.8: Transformation Settings for Georeferencing an Image with QGIS

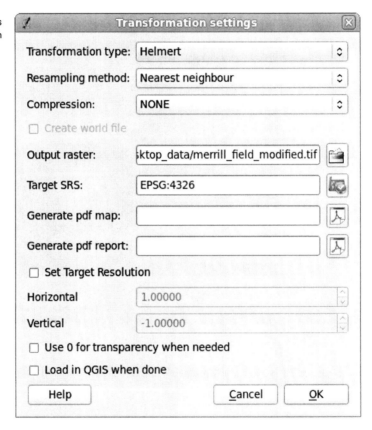

11

Projections and Coordinate Systems

If the world were flat, it would be a lot easier—at least on mapmakers. Unfortunately, that's not the case, so we're faced with the age-old problem of depicting features on a spheroid (that's the earth) on a flat piece of paper (or screen).

To solve this problem over the years, people have come up with the concept of map *projections*. The key thing to remember about projections is that none of them is perfect. You simply can't represent the entire earth (or even a small part of it) on a flat surface without some distortion. The amount of distortion varies with the projection. Many projections are quite good when used for a small or regional area. If you try to use the same projection for a larger area, the distortion increases.

Let's look at the main problems with squashing the earth onto a piece of paper. It's impossible for a projection to maintain an accurate portrayal of area, distance, shape, and direction all at once. For this reason, you'll find that some projections are more suited for your use than others. For example, if we're interested in measuring the areas of lakes we've digitized, we want a projection that is equal-area. This means that for any given location on the map, the measured area will be correct. If we are interested in measuring distances, we obviously need an equi-distant projection.

In choosing a map projection, we need to decide whether our focus is on shape, direction, area, or distance. Once we know that, we can choose an appropriate projection. Of course, sometimes you don't get a choice. You are forced to work in a particular projection for one reason or another. When Harrison wanted to display his bird sightings on the DRG, he needed to make sure they were in the same projection. Rather than "warp" the raster, he found it easier to convert his sightings from geographic to UTM, the same projection as the DRG. You can warp your rasters (no, it's not illegal) if you find it more convenient than transforming the dozens of vector layers in your dataset. For an example of warping a raster, see Section 13.2, *Raster Conversion*, on page 224.

There are plenty of books and online resources that delve into the details of projections and datums. Our goal here is to give you a brief yet practical introduction to provide what you need to know to work with your data. At the end of the chapter, you'll find some additional resources you can use to learn more about the sometimes complex world of projections and coordinate systems.

11.1   *Projection Flavors*

Projections come in three main flavors: planar or azimuthal, conic, and cylindrical. The type indicates how the projection is constructed.

Azimuthal
>   In an azimuthal projection, the sphere (that's the earth) is projected onto a flat or planar surface. Examples of azimuthal projections include Orthographic, Stereographic, Gnomonic, Azimuthal Equal Distant, and Lambert Azimuthal Equal Area.

Conic
>   In a conic projection, a spherical surface is projected on to a cone. Examples of conic projections include Albers Equal Area, Lambert Conformal, Equidistant Conic, and Polyconic Conic.

Cylindrical
>   In a cylindrical projection, the sphere is projected on to the walls

of a cylinder. Examples of cylindrical projections include Mercator, Transverse Mercator, Oblique Mercator, Space Oblique Mercator, and Miller Cylindrical. There are also a couple of pseudo-cylindrical projections: Robinson Pseudo-cylindrical and Sinusoidal Equal Area Pseudo-cylindrical.

Of course, the last projection we need to mention is Geographic, which really isn't a projection at all. It's just a coordinate system of latitude and longitude in a given datum. You'll find a lot of data in Geographic—just make sure that the datum matches your intended use.

What's a Datum?

Although we could go into a complicated definition, a *datum* is just a model for determining the coordinates of points on the earth. You are likely to encounter the North American Datum of 1927 (NAD 27), the North American Datum of 1983 (NAD 83), and the World Geodetic System of 1984 (WGS 84).

It's important to make sure that either your data is in the same datum or you are using software that can convert between datums on the fly. What happens if you mix datums? Your data won't line up as it should. All projections are based on a datum, so make sure to understand your data before you start trying to put it all together.

11.2   Working with Projections

Let's look at what we need to know to work with projections. When using a chunk of data in our OSGIS software, we should determine the following:

- Projection
- Units of measure
- Datum

In practice, most people don't care too much about these things until they have a problem and the data doesn't overlay properly. Or worse yet, it looks fine, and they think it's fine, but it's not. This can lead to making decisions based on bad information. So, it's best to check your projection parameters to make sure that everything is displayed where it belongs.

Harrison just acquired a new megasized bird layer and is excited to use it. Let's look at the ways he can discover the projection information.

Determining the Projection

There are a number of ways to determine the projection for a dataset. In some cases it's pretty easy—in others it can be quite difficult if the person creating the data failed to include the information from the outset. For example, a shapefile is usually (or should be) accompanied by a prj file containing the projection information. This is just a text file containing the projection parameters in what is known as Well-Known Text (WKT) format, defined by the OGC OpenGIS Simple Features Implementation Specification for SQL. If you open a prj file in your favorite text editor, you'll see something similar to this:

```
GEOGCS["WGS 84",DATUM
 ["WGS_1984",SPHEROID["WGS 84",6378137,298.257223563,AUTHORITY["EPSG","7030"]],
 AUTHORITY["EPSG","6326"]],
 PRIMEM["Greenwich",0,AUTHORITY["EPSG","8901"]],
 UNIT["degree",0.01745329251994328,AUTHORITY["EPSG","9122"]],
 AUTHORITY["EPSG","4326"]
]
```

From looking at the WKT, we can determine that this layer is in geographic coordinates (GEOGCS), meaning it's not projected. We also see that it is based on the WGS-84 datum and the units are degrees. There is also a bunch of authority information that indicates the EPSG codes for each section. See the sidebar on page 164 for more information on EPSG.

This gives us enough information to determine whether the layer can be used with the rest of our data or whether we need to do some conversion to make things line up properly. Let's look at the WKT for a projected coordinate system:

```
PROJCS["Albers Equal Area",
  GEOGCS["clark66",DATUM["D_North_American_1927",
    SPHEROID["clark66",6378206.4,294.9786982]],
    PRIMEM["Greenwich",0],
    UNIT["Degree",0.017453292519943295]],
  PROJECTION["Albers"],
  PARAMETER["standard_parallel_1",55],
  PARAMETER["standard_parallel_2",65],
  PARAMETER["latitude_of_origin",50],
  PARAMETER["central_meridian",-154],
  PARAMETER["false_easting",0],
  PARAMETER["false_northing",0],
  UNIT["Meter",1]
]
```

Here we see that the layer is projected (PROJCS); it's based on the Clarke 1866 spheroid, NAD 27 datum; the projection is Albers; and the units are meters. The WKT also contains the parameters (standard parallels, origin, and central merdian) for the Alaska Albers Equal Area Conic projection.

The key things to look for are the PROJCS or GEOGCS at the beginning of the WKT specification, the PROJECTION and DATUM keywords, and the UNIT keyword. These are enough to tell us whether it's suitable for use with our other data.

The second way to determine the projection for a layer is using the gdalinfo command for rasters and ogrinfo for vector layers. Recall that these utilities are part of GDAL/OGR. We took a look at these commands in Chapter 5, *Working with Vector Data*, on page 39 and Chapter 6, *Working with Raster Data*, on page 73. If you look back to those chapters, you'll see that both commands display the projection information in the same WKT format we just looked at.

The final way to determine the projection of a dataset is to load it into your OSGIS application and check the properties for the layer. This works fine and is handy if you already have the layer loaded, but it's quicker to just look at the WKT for a vector or use one of the GDAL/OGR commands to get the information. Since gdalinfo and ogrinfo work with nearly every format you'll encounter, it's worth installing and using them.

> **EPSG Notation**
>
> ---
>
> If you are wondering about the EPSG notation that has popped
> up in the Well-Known Text of coordinate systems, it represents
> a dataset of coordinate systems formerly distributed by the for-
> mer European Petroleum Survey Group. In 2005 the European
> Petroleum Survey Group was absorbed into the OGP Surveying
> and Positioning Committee. The OGP continues to distribute
> the dataset.
>
> The Geodetic Parameter Set contains a unique code for each
> coordinate system, as well as details about the projection. Many
> OSGIS applications can use the EPSG code as input when doing
> transformations. If you have PROJ.4 installed, you should have
> a copy of the epsg file on your system. This file is a simplified
> subset of the EPSG definitions and maps the EPSG number to
> the corresponding PROJ.4 parameters. The full EPSG database
> and documentation is available from OGP at www.epsg.org

Data Problems

If you find that your data isn't lining up like you expect, it's either
a projection problem or you just have lousy data. Seriously, though,
most alignment problems are due to either a projection problem or
differing datums. The first thing to do is use your sleuthing skills to
examine the projection and datum for your layers. If you are seeing
a big difference in alignment, it's likely a projection problem. If it's
a small difference (less than 500 meters, for example), you likely
have a datum problem.

To get to the bottom of it, you have a couple of choices. If your
OSGIS application supports on-the-fly projection (and datum shift
if required), enable it and make sure that the projections are recognized by the application. If this solves the problem, you don't need
to do anything else; just keep in mind that you are transforming
data on the fly—you haven't changed the original data in any way.

If on-the-fly transformation isn't an option, you will have to manually transform the data to get it into a projection you can use. If
you are using PostGIS, you can create a SQL view of your data that

transforms the geometries using the OGC `transform` function. You
would then load the view into QGIS or your other application, and
the transformation would be done automatically, at the expense of
a bit of performance. For example, if we have a `towns` layer in our
PostGIS database that we would like to use with another layer that
is geographic, we can create a view to do the job.[1] First let's look at
the schema of the table:

[1] The coordinate system for our towns layer in PostGIS is Albers Equal Area.

```
gis_data=# \d towns
                                        Table "public.towns"
   Column   |         Type          |                Modifiers
------------+-----------------------+-----------------------------------------
 gid        | integer               | not null default
            |                       | nextval('towns_gid_seq'::regclass)
 name       | character varying(20) |
 class      | character varying(35) |
 pop        | integer               |
 shape      | geometry              |
 ...
gis_data=#
```

Next we'll select a few records from the table to examine the coordinates, just so we can see that our transform works when we're all
done:

```
gis_data=# SELECT gid, ASTEXT(shape) AS coordinates FROM towns LIMIT 5;
 gid |            coordinates
-----+----------------------------------
   1 | POINT(-820805.1875 506479.46875)
   2 | POINT(-384408.6875 1333234.375)
   3 | POINT(88849.421875 881200.0625)
   4 | POINT(-565926.875 1174128)
   5 | POINT(-307637.65625 1451385.5)
(5 rows)
```

Now we can create the view using the `transform` function to convert from the projected coordinate system to WGS 84 geographic
(EPSG:4326). We know the EPSG code is 4326 because we looked it
up using one of the methods that you'll learn about in just a moment. We can now create the view:

```
gis_data=# CREATE VIEW towns_geo AS SELECT gid, name, class, pop,
           TRANSFORM(shape,4326) AS shape FROM towns;
CREATE VIEW
gis_data=#
```

Now we can use our towns_geo view as a layer in QGIS. Just to make sure the transform works, let's select a few records to see whether the coordinates look like they are geographic:

```
gis_data=# SELECT gid, ASTEXT(shape) AS coordinates FROM towns_geo LIMIT 5;
 gid |                  coordinates
-----+-----------------------------------------------
   1 | POINT(-157.571463607738 51.336408724444)
   2 | POINT(-155.770900879198 53.6566126757253)
   3 | POINT(-153.605243492496 52.4282023272413)
   4 | POINT(-156.577035823078 53.203783217787)
   5 | POINT(-155.42947766785 53.9856939176996)
(5 rows)
gis_data=#
```

Sure enough, our view is returning coordinates in WGS 84 geographic. The view gives us a quick way to transform our data on the fly and be able to visualize it with our other data. Our original data isn't changed—we're just doing a transform on the fly.

If it comes down to it, you can always transform your data, creating a new dataset in the appropriate projection. We'll cover transformation for both vector and raster layers in more detail later when we look at using command-line utilities. If you're curious, see Section 13.2, *Coordinate System Conversion*, on page 222 and Section 13.2, *Raster Conversion*, on page 224.

11.3   The PROJ.4 Projections Library

PROJ.4 is a cartographic projections library that is used in many, if not most, open source GIS applications, both on the desktop and on the Internet. It was originally developed by the USGS and is now maintained by a group of volunteers.

You may be wondering why we are mentioning a library—turns out that PROJ.4 also comes with some handy utilities for experimenting with projections and doing interactive transformations:

proj and invproj

   Performs forward and inverse transformations for a large number of projections. With your projection parameters in hand, you can perform a forward calculation using proj (geographic to projected) or an inverse calculation using `invproj` (projected to geographic). Neither of these utilities does datum shifts.

cs2cs

Performs transformations between coordinate systems, including datum shifts. With `cs2cs`, you supply the parameters for both coordinate systems, specifying which is the target or "to" system.

geod and invgeod

Performs forward and inverse Great Circle (geodesic) transformations. This allows you to calculate latitude, longitude, and back azimuth given a starting point, azimuth, and distance. You can also determine the azimuths (forward and back) and distance between two known points.

nad2nad

Performs datum conversions between the North American 1927 and 1983 datums. The same conversions can be accomplished using `cs2cs`.

Let's look at an example. Say Harrison has loaded up some of his bird data in his favorite desktop application and it contains a DRG. If you remember correctly, DRGs come in a UTM projection. Harrison is curious about a couple of bird observations on his map. When he moves his cursor over the points, he sees big numbers for the coordinates in the status bar of his application. He knows his DRG is in UTM Zone 6, NAD27 datum, but he wants to know the approximate geographic coordinates for the point.[2] PROJ.4 can quickly answer his question. First he has to know the parameters for the UTM projection in order to do the conversion. There are several ways to do this—perhaps the easiest is to look up the `proj` string in the `epsg` file that is installed with PROJ.4.[3] On a Linux system you'll find this in `/usr/share/proj/epsg`. To locate the projection, you can open the `epsg` file in your favorite text editor and search, or you can use `grep` from the command line:

[2] If you think there is more than one way to do this, you are right—depending on the software you are using.

[3] The epsg file included with PROJ.4 does not include a number of datums outside the United States and Canada. You can find a complete list of datums in your GRASS install in etc/datum.table.

```
$ grep -C 1 -i "utm zone 6n" /usr/share/proj/epsg| \
    grep -i "nad27"
<26705> +proj=utm +zone=5 +ellps=clrk66 +datum=NAD27 +units=m +no_defs  <>
# NAD27 / UTM zone 6N
<26706> +proj=utm +zone=6 +ellps=clrk66 +datum=NAD27 +units=m +no_defs  <>
```

Here we used `grep` to search for "utm zone 6n" and told it to print one line on either side of the match (`-C 1`) and ignore the case (`-i`). Since we wanted only NAD 27 projections, we piped the output to `grep` again to show only those lines containing "nad27." From the result we get a couple of things: the EPSG number, in this case 26706, and the string that we need to use with `proj`. If you can't find the `epsg` file on your system, you can search for your projection using the tools on the Spatial Reference website.[4] Entering "nad27 utm zone 6n" as the search string will quickly find the projection. You can then copy the `proj` parameters from the website.

Now that Harrison has the projection parameters, he can convert his coordinates from UTM Zone 6N to geographic using `invproj`:

```
$ invproj +proj=utm +zone=6 +ellps=clrk66 +datum=NAD27 \
  +units=m +no_defs
312244.49 6795460.41
150d30'W        61d15'N
```

Harrison entered the coordinates he read from his status bar (312244.49, 6795460.41) and got the results in degrees and minutes (150d30'W, 61d15'N). Lucky for Harrison, his birds like to roost on nice clean coordinates. Just to convince himself that this works, he plugs the answer back in to the `proj` command to see whether he gets his original UTM coordinates:

```
$ proj +proj=utm +zone=6 +ellps=clrk66 +datum=NAD27 \
  +units=m +no_defs
150d30'W  61d15'N
312244.49        6795460.41
```

Sure enough it worked. He could have specified the latitude and longitude using decimal degrees (-150.5 61.25) and gotten the same result. PROJ.4 also has a number of other options, including the ability to customize the output format to your taste. You can see that PROJ.4 can be useful for doing interactive transforms.

[4] http://spatialreference.org

If you wanted to transform from a UTM projection to an Albers
Equal Area, you would have to do an inverse (invproj) to get the
latitude and longitude for the UTM point and then do a forward
projection (proj) using the Albers parameters to get the final result.
The cs2cs program simplifies this by allowing you to specify both
coordinate systems. Let's convert our UTM point from the previous
example to Alaska Albers coordinates using cs2cs:

```
$ cs2cs  +proj=utm +zone=6 +ellps=clrk66 +datum=NAD27 \
  +units=m +to +proj=aea +lat_1=55 +lat_2=65 +lat_0=50 +lon_0=-154 \
  +x_0=0 +y_0=0 +datum=NAD27 +units=m +no_defs
312244.49 6795460.41
187115.08       1257043.95 0.00
```

Now we have Albers coordinates 187115.08, 1257043.95 as a result
of our transformation. You are probably wondering what the 0.00
means. This is height above (or below) the ellipsoid. Since both
datums were NAD27 and they are both based on the same ellipsoid,
there is no difference. Let's do a datum shift to illustrate how it's
done and compare the results:

```
$ cs2cs +proj=utm +zone=6 +datum=NAD27 +units=m \
  +to +proj=aea +lat_1=55 +lat_2=65 +lat_0=50 +lon_0=-154 +x_0=0 +y_0=0 \
  +datum=NAD83 +units=m +no_defs
312244.49 6795460.41
186991.95       1256960.61 0.00
```

Here we see the result, which is a bit different from the NAD27
result. The total shift in coordinates is approximately 184 meters. If
you don't believe it, break out your high-school math book and use
the distance formula (Pythagorean Theorem) to check the result.

If you want to transform a lot of points, you can provide the coordinates to cs2cs from a file to do batch conversions. See the documentation for cs2cs for details on the available options.

You may be wondering which of the PROJ.4 utilities you should use.
In general, cs2cs is your best bet since it supports transformations
between projected coordinate system as well as datum transforms.
If you just need to go to/from a projected coordinate system to geographic without a datum transform, the proj and invproj utilities

do an acceptable job.

11.4 More Resources

The U.S. Geological Survey has a "poster" that provides an excellent overview of projections and the characteristic of each. The poster is available for download free of charge.[5]

[5] Download it from http://erg.usgs.gov/isb/pubs/MapProjections/projections.pdf.

Published in 1987, the USGS Professional Paper, "Map Projections: A Working Manual" provides a good overview of projections and includes the mathematical formulae needed to transform more than twenty-five different coordinate systems. The paper is available online.[6]

[6] http://pubs.er.usgs.gov/usgspubs/pp/pp1395

You might find it hard to believe, but an animated movie made in 1947 provides an excellent beginner's introduction to the problems of portraying a spherical earth on a flat surface. Entitled *The Impossible Map*, the movie is available online from the National Film Board of Canada.[7]

[7] http://www3.nfb.ca/animation/objanim/en/films/film.php?sort=title&id=14229

The American Society for Photogrammetry & Remote Sensing maintains a repository of the "Grids and Datums" column from each issue of its journal.[8] Datums and grids for a number of regions around the world are documented in each column since 1998.

[8] http://www.asprs.org/persjournals/PE-RS-Journals/Grids-Datums.html

Lastly, using your favorite search engine on the term *map projections* will bring up enough reading material to keep you busy for a while. Becoming an expert on projections takes some time and effort— learning enough to be proficient with your data is as simple as being able to identify what you have and making your software work for you.

12

Geoprocessing

It's often not enough to have data and look at it. We almost always want to do some sort of manipulation or processing. This is where geoprocessing comes in. We'll use the broad definition of geoprocessing to include any kind of data manipulation and analysis. To some extent, you could consider importing data as a geoprocessing operation. In this chapter, we'll look at some other operations to include the following:

- Projection
- Line-of-sight analysis
- Watershed modeling
- Hillshading
- Clipping features
- More complex importing
- Grid algebra

If you are wondering what tools are available in the OSGIS stack to accomplish these tasks, the answer is: most of them. Support varies, however, QGIS, gvSIG, and JGrass all provide you with some geoprocessing capability. By far the widest range of geoprocessing capability is provided by GRASS. If you plan to do advanced geoprocessing, you're likely going to need GRASS. The goal in this chapter is not to instruct you in the use of all the GRASS geoprocessing tools but rather to introduce you by way of example to the possibilities.

12.1 Projecting Data

That's right—we're going to take one more look at projecting data, even though we've given it a good going over in Chapter 11, *Projections and Coordinate Systems*, on page 159. Once again, depending on how you view and use your GIS data, you may find you need to project the data to another coordinate system to make your life easier, particularly when doing analysis. Some of the desktop GIS applications support "on-the-fly" projection of data. This means you set the default projection for the map, and every layer that is added is reprojected to that coordinate system, assuming it's different from the default. This is a two-edged sword. On one hand, it's very convenient. On the other, there is a performance penalty in that every point or vertex must be transformed as the features are drawn. The size of the penalty depends on a number of factors and really can't be generalized; however, suffice it to say that as long as you're transforming coordinates on the fly, it's going to be slower.

As we've seen, the solution is to transform the data to a coordinate system that is suitable for your project work. In Section 13.2, *Coordinate System Conversion*, on page 222, you'll see how to use ogr2ogr to transform OGR-supported layers. In this section, we'll look at how to change the coordinate system of our data using GRASS.

GRASS has two commands that are used to project data: r.proj for raster maps and v.proj for vector maps. The key to projecting data is having your locations properly defined and set up. If you need a hand getting started with GRASS locations and mapsets, take a look at Section 18.1, *Location, Location, Location*, on page 323.

When you project a map, it ends up in your *current* location and mapset. You could think of it as copying the map from its original location to your current one, transforming the coordinates as it goes. You don't specify any projection parameters when projecting GRASS maps, since it all hinges on the locations involved and locations always have a projection/coordinate system defined.

Let's look at the usage for projecting a vector map using v.proj:

```
GRASS 6.4.2RC1 (alaska):~/geospatial_desktop_data > v.proj help

Description:
 Re-projects a vector map from one location to the current location.

Keywords:
 vector, projection, transformation

Usage:
 v.proj [-lz] [input=name] location=name [mapset=name] [dbase=path]
   [output=name] [--overwrite] [--verbose] [--quiet]

Flags:
  -l   List vector maps in input location and exit
  -z   Assume z co-ordinate is ellipsoidal height and transform if possible
       3D vector maps only
 --o   Allow output files to overwrite existing files
 --v   Verbose module output
 --q   Quiet module output

Parameters:
     input    Name of input vector map
  location    Location containing input vector map
    mapset    Mapset containing input vector map
     dbase    Path to GRASS database of input location
    output    Name for output vector map (default: input)
```

As you can see, there aren't many required parameters. You have to know where the map resides, including the full path to the GRASS database if it's in a different database than the target location. Usually you get away with just entering the map name and location and specifying the output name. As usual, GRASS gives you an --overwrite option in case you need to run the command more than once to get it right. For example, if we have a lakes map in a location named alaska_albers, we can project it to a UTM Zone 6 location, alaska_utm6_nad27 using v.proj[1] We begin by starting GRASS in the location where we want the projected map to reside, in this case our UTM location, and then proceeding:

[1] Projecting a region the size of Alaska into a single UTM zone is not really appropriate, but it illustrates the process of projecting vectors with GRASS.

```
GRASS 6.4.2RC1 (alaska_utm6_nad27):~ > v.proj input=lakes \
  location=alaska_albers mapset=PERMANENT output=lakes

Input Projection Parameters: +proj=aea +lat_1=55 +lat_2=65 +lat_0=50
+lon_0=-154 +x_0=0 +y_0=0 +no_defs +a=6378206.4 +rf=294.9786982
+nadgrids=/usr/local/grass-6.3.cvs/etc/nad/alaska
Input Unit Factor: 1

Output Projection Parameters: +proj=utm +zone=6 +a=6378206.4
+rf=294.9786982 +no_defs
+nadgrids=/usr/local/grass-6.3.cvs/etc/nad/alaska
Output Unit Factor: 1
Re-projecting vector map...
Building topology ...
177 primitives registered
Building areas:  100%
87 areas built
87 isles built
Attaching islands:  100%
Attaching centroids:  100%
Topology was built.
Number of nodes     :    178
Number of primitives:    177
Number of points    :    0
Number of lines     :    0
Number of boundaries:    89
Number of centroids :    88
Number of areas     :    87
Number of isles     :    87
Number of incorrect boundaries   :   2
Number of duplicate centroids    :   1
GRASS 6.4.2RC1 (alaska_utm6_nad27):~ >
```

We specified the name, location, and mapset for the map (layer) we wanted to project and specified the output name to be lakes. GRASS gives us a whole bunch of information as it proceeds with the projection process. Once complete, we have our new lakes layer projected to UTM Zone 6—we'll use v.info to see what we created:

```
GRASS 6.4.2RC1 (alaska_utm6_nad27):~ > v.info map=lakes
+------------------------------------------------------------------------+
| Layer:    lakes                                                        |
| Mapset:   gsherman                                                     |
| Location: alaska_utm6_nad27                                            |
| Database: /home/gsherman/grassdata                                     |
| Title:                                                                 |
| Map Scale:  1:1                                                        |
| Map format: native                                                     |
| Name of creator: gsherman                                              |
| Organization:                                                          |
| Source date:    Fri Nov  4 21:04:24 2011                               |
|------------------------------------------------------------------------|
|   Type of Map:  Vector (level: 2)                                      |
|                                                                        |
|   Number of points:      0        Number of areas:    87               |
|   Number of lines:       0        Number of islands:  87               |
|   Number of boundaries:  89       Number of faces:    0                |
|   Number of centroids:   88       Number of kernels:  0                |
|                                                                        |
|   Map is 3D:             0                                             |
|   Number of dblinks:     1                                             |
|                                                                        |
|   Projection: UTM (zone 6)                                             |
|         N: 12862988.157    S: 0.000                                    |
|         E: 420703.718    W: -2476603.714                               |
|         B: 0.000         T: 0.000                                      |
|                                                                        |
|   Digitize threshold: 0                                                |
|   Comments:                                                            |
|                                                                        |
+------------------------------------------------------------------------+
```

You can see once you have your locations set up in GRASS, projecting to a new coordinate system is fairly simple. Using the same methodology, you can also project rasters using r.proj, which we'll do as part of our next geoprocessing task. Let's move on to some topics that are more along the lines of "classic" geoprocessing.

12.2 Line-of-Sight Analysis

Line-of-sight analysis (LOS) is interesting from a curiosity standpoint as well as for hard analysis. Suppose you want to know what you can see from the top of the local mountain (assuming you don't live in Kansas). With the right data to work with, LOS analysis can show you all the areas that are visible from a given point on the map. Some practical applications are determining the visibility of features in site planning. Can the new garbage dump be seen from the local park? How many people will be able to see the new 75-foot-tall monster transmission tower (I have a new one in my backyard)? When doing LOS analysis, we can specify not only the location to view from but also the height of the observer (that would be us). Let's take a look at a simple LOS example.

In our example, we will use the GRASS r.los command to create a viewshed (area we can see) from a given point. To do the analysis, we need a raster dataset that has elevation information. A couple of examples are the USGS Digital Elevation Model (DEM) product and the National Elevation Dataset (NED).[2]

We will use a 1:63,360 DEM (that's 1 inch = 1 mile) in our LOS analysis. The steps to get from a raw DEM to our LOS are as follows:

1. Download the DEM.
2. Import the DEM into our world latitude-longitude mapset.
3. Project the DEM into the Albers coordinate system.
4. Use r.los to do the LOS analysis.
5. Display the results.

Getting the DEM

The first task is to get the DEM and get it ready for use in the analysis. The DEM we chose is for the Anchorage C6 quadrangle in Alaska.[3] The file (ancc6.gz) came gzipped, so before it can be imported, it must be unzipped. We used gzip -d to unzip it and then renamed it to ancc6.dem. If we wanted to, we could view it right now using QGIS because it supports USGS ASCII DEMs. To import it into GRASS, we need a geographic location since the DEMs coordinates are in degrees of latitude and longitude. The datum of the location must match the DEM as well. In the case of our DEM, that's NAD 27. If you can't remember the gory details of creating a new GRASS location, refer to Section 18.1, *Location, Location, Location*, on page 323.

[3] You can download this DEM at http://agdc.usgs.gov/data/usgs/geodata/dem/63K/demlist_A.html.

To import the DEM into our geographic location, from the GRASS shell we use the following:

```
r.in.gdal input=ancc6.dem output=ancc6_dem title="Anchorage C6 DEM"
```

Now we need to project the DEM to our Albers coordinate system. You might be asking why we have to project it. Well, the answer is, r.los doesn't work with geographic coordinates. If you try it, you'll get a nice message along the lines of this:

```
ERROR: Lat/Long support is not (yet) implemented for this module.
```

To project the DEM, we use the r.proj command. But first we need to have an Alaska Albers location created using the proper parameters. For this example, we created one and set the default region (part of the creation process) to just the area of our DEM. If you need to know the parameters for a location, you can use g.proj -p from the GRASS shell to print the projection information for the current location.

```
GRASS 6.4.2RC1 (alaska_albers):~ > g.proj -p
-PROJ_INFO------------------------------------------------
name       : Albers Equal Area
proj       : aea
datum      : nad27
ellps      : clark66
lat_1      : 55
lat_2      : 65
lat_0      : 50
lon_0      : -154
x_0        : 0
y_0        : 0
towgs84    : 0,0,0,0,0,0,0
no_defs    : defined
-PROJ_UNITS----------------------------------------------
unit       : Meter
units      : Meters
meters     : 1
```

To project the DEM, we start GRASS in the target location (in this case Alaska Albers) and use the r.proj command:

```
r.proj input=ancc6_dem location=geopgraphic_nad27 output=ancc6_dem
```

Notice we specify the source location where the geographic version

of the DEM resides. Now we have the DEM ready to use in our
analysis.

Doing the Line-of-Sight Analysis

Now that the data is in order, we can get down to doing some
analysis. To use `r.los`, we need to know where the observer is
located in map coordinates, as well as the maximum distance we
want the analysis to consider. We can also apply a height to the
observer if we desire. For now, we'll do a simple analysis using
a point located on a gravel bar in the river bottom to determine
what we can see. To get the coordinates for the observer (that's us
standing on the gravel bar), we just used QGIS or GRASS to get the
location of the mouse cursor in map units. With that, we can run
`r.los`:

```
r.los input=ancc6_dem output=los_river coordinate=259315,1307037 max_dist=3000
```

This gives us a line-of-site analysis extending 3,000 meters from our
location. In Figure 12.1, on the facing page, you can see the results
of the analysis, with those areas that we can see from our position
(the red circle) shown in light blue. We've overlaid the LOS results
on our DRG so we can compare the analysis with the topography.
As you might expect, our line of sight is somewhat limited when
standing in the river bottom. We can't see very far to the west,
basically just along the top of the river bluff. We can see upstream
and downstream a fair bit, as well as to the east, which is on the
inside of the river bend and consequently doesn't have a high bank.

To test the LOS ability of GRASS, we took a simple example from
the river bottom. We hope that was enough for you to see the power
of this type of analysis. We'll continue our river theme in the next
section by looking at the watershed modeling tools in GRASS.

12.3   Hydrologic Modeling

GRASS includes a number of modules for hydrologic modeling, including modules to create and analyze watershed basins, carve out

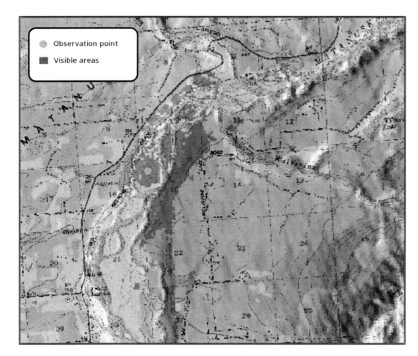

Figure 12.1: Results of line-of-sight analysis in GRASS

streams in a DEM, trace a drainage path, simulate flooding, and perform a host of other functions. These modules provide a sophisticated toolset for your hydrologic modeling needs.

To illustrate one of these tools, we'll take a simple example and raise the sea level by 100 meters using the ancc6_dem DEM. The r.lake module allows you to create a new raster map portraying the filling of an area on a DEM to a given level. You just specify the start point and the water level. You don't have to be real picky about the start point because GRASS analyzes the DEM and fills it such that the deepest point will be equal to the depth you chose. This means you can actually pick what will become a very shallow area (in other words, a higher elevation), and the lake will be created properly. After all, water does flow downhill.

First let's look at the usage for r.lake:

```
GRASS 6.4.2RC1 (alaska_albers):~ > r.lake help

Description:
 Fills lake from seed at given level.

Keywords:
 raster, hydrology

Usage:
 r.lake [-no] dem=name wl=value [lake=name] [xy=east,north]
   [seed=name] [--overwrite] [--verbose] [--quiet]

Flags:
  -n   Use negative depth values for lake raster map
  -o   Overwrite seed map with result (lake) map
  --o   Allow output files to overwrite existing files
  --v   Verbose module output
  --q   Quiet module output

Parameters:
   dem   Name of terrain raster map (DEM)
    wl   Water level
  lake   Name for output raster map with lake
    xy   Seed point coordinates
  seed   Name of raster map with seed (at least 1 cell > 0)
```

The command is pretty straightforward. Notice that we can choose
to assign negative values to the lake map using the -n option. This
means that if we query a given cell, the value will be negative, indicating a depth from the surface of the lake.

The seed coordinates specify the starting point of the calculations.
We could use a raster map as a seed as long as it has one cell with
a value greater than zero. Why would we want to do this? If we
wanted to create a series of maps showing an increasing water level,
we could use the previous output as the seed for the next map.
Another important point is that the water level must be specified in
DEM units—in our case, meters.

To flood our DEM, we pick a point in the southwest corner somewhere and use the following command:

```
r.lake dem=ancc6_dem lake=ancc6_lake_100m xy=258686.903427,1298819.69314 wl=100
```

This creates a new map named ancc6_lake_100m, as shown in Figure 12.2, on the next page. Each flooded cell in the raster has a

value indicating the depth, while those that are above water are set to null. This means the underlying layer(s) on our map are visible so we can see what land remains.

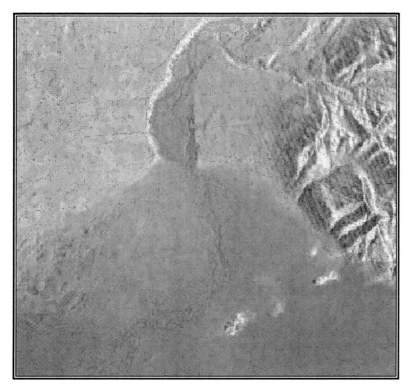

Figure 12.2: Raising sea level by 100 meters

If you look carefully at the newly created lake, you can see the original river course underneath. It flows from the top center of the map down and then to the west. This was the extent of the water before we flooded the area, apart from a few lakes in the southwest quadrant of the map that are now completely underwater. You can see from the result that raising sea level 100 meters isn't a good thing. We flooded several lakes, along with a bunch of subdivisions and a town or two.

Using r.lake is a simple example of a pretty powerful tool in the GRASS hydrologic modeling toolbox. If you use your imagination, you could combine this tool with a bit of shell script to loop

through multiple iterations of rising sea level, saving each image using r.out.mpeg to create an animation. But we'll leave that exercise
to you.

12.4   Creating Hillshades

You no doubt have seen those fancy shaded relief maps. Now we
are going to see how to create one from a DEM using the GRASS
r.shaded.relief module. Again we'll use the ancc6_dem DEM as
the starting point. First let's get a look at the usage and options for
r.shaded.relief.

```
GRASS 6.4.2RC1 (alaska_albers):~ > r.shaded.relief help

Description:
 Creates shaded relief map from an elevation map (DEM).

Keywords:
 raster, elevation

Usage:
 r.shaded.relief map=string [shadedmap=string] [altitude=value]
   [azimuth=value] [zmult=value] [scale=value] [units=string]
   [--overwrite] [--verbose] [--quiet]

Flags:
 --o   Allow output files to overwrite existing files
 --v   Verbose module output
 --q   Quiet module output

Parameters:
        map   Input elevation map
  shadedmap   Output shaded relief map name
   altitude   Altitude of the sun in degrees above the horizon
              options: 0-90
              default: 30
    azimuth   Azimuth of the sun in degrees to the east of north
              options: 0-360
              default: 270
      zmult   Factor for exaggerating relief
              default: 1
      scale   Scale factor for converting horizontal units to elevation units
              default: 1
      units   Set scaling factor (applies to lat./long. locations only, none: scale=1
              options: none,meters,feet
              default: none
```

The command is pretty straightforward and has several options for creating the shaded relief map from the DEM, including sun angle and altitude. We can also exaggerate the relief to get a more dramatic effect. We'll start simple and create a standard shaded relief map:

```
GRASS 6.4.2RC1 (alaska_albers):~ > r.shaded.relief map=ancc6_dem \
  shadedmap=ancc6_shade1
Calculating shading, please stand by.
 100%
Color table for raster map <ancc6_shade1> set to 'grey'
Shaded relief map created and named <ancc6_shade1>.
```

This map is all default settings. Generally, the default light settings (altitude and azimuth) produce good results, unless you have particular needs. Notice that when we created the hillshade, it set the color to gray. This is because r.shaded.relief is actually a shell script that runs both r.mapcalc and r.colors to create the hillshade. Let's try one with some vertical exaggeration, and then we'll compare:

```
GRASS 6.4.2RC1 (alaska_albers):~ > r.shaded.relief map=ancc6_dem \
  shadedmap=ancc6_shade2 zmult=4
Calculating shading, please stand by.
 100%
Color table for raster map <ancc6_shade2> set to 'grey'
Shaded relief map created and named <ancc6_shade2>.
```

Now we have an "out-of-the-box" hillshade and one with exaggerated relief (four times). In Figure 12.3, on the following page, you can see the two side by side, with the default hillshade on the left and the 4X exaggeration on the right. I'll leave it up to you to decide which looks better.

Colorizing the Hillshade

A gray hillshade is nice but a bit boring. Let's look at how to make our hillshade nicely colored and even export it to a georeferenced TIFF. The process consists of two steps: colorizing the DEM and combining it with the hillshade to make the final product.

To begin, we will create a rules file to define colors by percentage

Figure 12.3: Hillshade with no exaggeration (left) and 4X exaggeration (right)

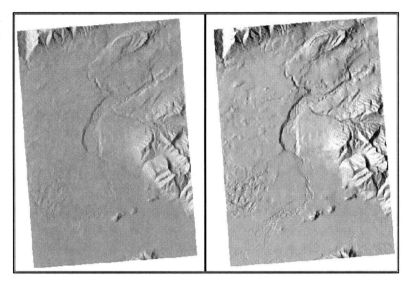

of the range of elevations in the raster. The rules file specifies percentages and a color in RGB notation and is taken directly from the manual page for r.colors.

```
0% 0 230 0
20% 0 160 0
35% 50 130 0
55% 120 100 30
75% 120 130 40
90% 170 160 50
100% 255 255 100
```

We save this as myelevation.rules and will use it in a minute. To create the colored hillshade, we set the colors for the DEM to those in the rules file we just built and then do some magic with the r.his and r.composite commands. Here is a script that does the whole process for us:

create_hillshade.sh

```
1  #!/bin/sh
2  r.shaded.relief map=ancc6_dem shadedmap=ancc6_shade zmult=4 --overwrite
3  cat myelevation.rules |r.colors map=ancc6_dem color=rules
4  r.his -n h_map=ancc6_dem i_map=ancc6_shade r_map=ancc6_r \
5    g_map=ancc6_g b_map=ancc6_b --overwrite
6  r.composite -d red=ancc6_r blue=ancc6_b green=ancc6_g \
7    output=ancc6_comb --overwrite
```

Let's take a look at each line of the script to see what it does. Line 2 creates the shaded relief map from the DEM, with a vertical exaggeration of 4. We specified the - -overwrite option (you can also use - -o) to allow us to run the script multiple times if need be, replacing the existing shaded relief map each time.

On line 3, we apply the color rules to the DEM by piping the contents of the rules file to the r.colors command. Now for the tricky part.

Next we use r.his to create red, green, and blue maps from the DEM and shaded relief as shown on line 4. The hue is taken from the input DEM and specified with the h_map parameter. This sets the color for each cell. The intensity or brightness of each cell is set from the shaded relief map using the i_map parameter. The remainder of line 4 contains the names for the output of the red, green, and blue maps.

Finally on line 6, we put the RGB maps back together with r.composite to create the final map named ancc6_comb. Now we have a nicely colored hillshade, as shown in Figure 12.4, on the next page.

Full color illustrations are available at http://geospatialdesktop.com.

You may be wondering about the value of the composite map we just created. Apart from displaying it in GRASS or QGIS, we can export it to other formats supported by the r.out.* suite of commands in GRASS. This includes creating a georeferenced TIFF that can be used with other GIS software for those poor folks not using GRASS and/or QGIS. Let's export the colored hillshade to a georeferenced TIFF that we can share:

```
r.out.tiff -t input=ancc6_comb output=ancc6.tif compression=packbit
```

The -t switch tells r.out.tiff to create a world file along with the TIFF. If we use gdalinfo on the newly created file, we can see exactly what we created.

Figure 12.4: Colored shaded relief
map created with GRASS

```
> gdalinfo ancc6.tif
Driver: GTiff/GeoTIFF
Size is 410, 551
Coordinate System is `'
Origin = (249831.465000,1318446.415363)
Pixel Size = (53.32700000,-53.30280132)
Image Structure Metadata:
  COMPRESSION=PACKBITS
Corner Coordinates:
Upper Left  (  249831.465, 1318446.415)
Lower Left  (  249831.465, 1289076.572)
Upper Right (  271695.535, 1318446.415)
Lower Right (  271695.535, 1289076.572)
Center      (  260763.500, 1303761.494)
Band 1 Block=410x6 Type=Byte, ColorInterp=Red
Band 2 Block=410x6 Type=Byte, ColorInterp=Green
Band 3 Block=410x6 Type=Byte, ColorInterp=Blue
```

We can see from the gdalinfo output that our new TIFF is a three-
band image and compressed with packbits compression. That's
good, because it's what we specified when we created the image.

Notice what's missing—there is no coordinate system defined for the image. When you export an image using r.out.tiff, it doesn't encode the coordinate system information into the TIFF. We could add this using gdal_translate. For more information on using gdal_translate and friends, see Section 13.2, *Using GDAL and OGR*, on page 212.

When you go to share your georeferenced hillshade maps with the rest of the world, make sure to include the world file. Otherwise, it may fall off the face of the earth when your friends attempt to display it with the rest of their data.

> ## Exporting Rasters from GRASS
>
> For the colorized hillshade, we used r.out.tiff to create a composite image from each of the raster maps (red, green, and blue) created by r.his. Compositing the maps results in some reduction of color, although this likely won't be noticeable to the human eye.
> For exporting single band rasters, r.out.gdal is a better choice. This is because a three-band image is always created by r.out.tiff, even though you specify a single GRASS raster as input.

12.5 *Merging Digital Elevation Models*

In this section we'll look at how to merge DEMs to create a single map layer in GRASS, and there is a very good reason to do so. As you look around the Internet, you'll find that a lot of the available data (DEMs included) is tiled. This means you have to download more than one file to get the complete dataset. Sometimes just having one tile is fine, as long as it covers the area you need. Other times, you might find you need several adjacent tiles to get the coverage you want. To illustrate the process, we'll merge several GTOPO30[4] DEMs into a single layer.

To begin, we fetched all the DEMs for the Americas and stashed them in a directory. The DEM files are distributed in a tar-gzipped

[4] http://eros.usgs.gov/#/ Find_Data/Products_and_Data_ Available/GTOPO30

format so you'll have to unpack them before proceeding. On Linux or OS X you can just use this:

```
tar -xzf w100s10.tar.gz
```

On Windows you can use a zip file manager that supports `tar.gz` files such as 7-zip. Before we proceed with the import, we need to edit the `HDR` file for each DEM, adding a line containing "PIXEL-TYPE SIGNEDINT." This ensures that the DEMs will be imported correctly (for more information, see the `r.in.gdal` manual page). Once we have the DEMs all unpacked and the `HDR` files properly edited, we can import them into our `world_lat_lon` location in GRASS using `r.in.gdal`:

```
r.in.gdal -e input=./gtopo30/W060N40.DEM output=w060n40
r.in.gdal -e input=./gtopo30/W060N90.DEM output=w060n90
r.in.gdal -e input=./gtopo30/W060S10.DEM output=w060s10
r.in.gdal -e input=./gtopo30/W100N40.DEM output=w100n40
r.in.gdal -e input=./gtopo30/W100N90.DEM output=w100n90
r.in.gdal -e input=./gtopo30/W100S10.DEM output=w100s10
r.in.gdal -e input=./gtopo30/W140N40.DEM output=w140n40
r.in.gdal -e input=./gtopo30/W140N90.DEM output=w140n90
r.in.gdal -e input=./gtopo30/W180N90.DEM output=w180n90
```

We could use the DEMs as is, loading each into GRASS or QGIS for display purposes. However, if we want to do some analysis or even create a combined shaded relief map, we need to put them all together. To do this, we use the GRASS `r.patch` command.

For your GeoTIFF and other file-based rasters you can create mosiacs using `gdal_merge.py`.

The usage for `r.patch` is pretty simple. All you do is provide a list of input DEMs and a name for the output. There are a couple of caveats, though. First, of course, you have to have a GRASS location for the area of interest, and second, make sure you set your GRASS region to the area covered by the combined DEMs. You can set the region using `g.region`. To "patch" the DEMs together, we use the following:

```
r.patch input=w060n40,w060n90,w060s10,w100n40,w100n90,w100s10, \
  w140n40,w140n90,w180n90 output=americas_dem
```

We just created a merged DEM named `americas_dem`, consisting of nine input DEMs. In Figure 12.5, on the next page, you can see the result, with the color map set to the same we used in Section 12.4,

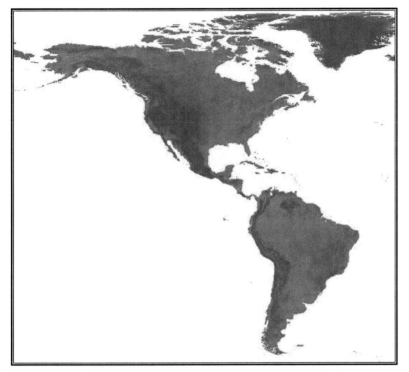

Figure 12.5: Merged GTOPO30 DEM

Colorizing the Hillshade, on page 183. We could now create a nice hillshade from the DEM or use it in some sort of analysis, such as melting the Antarctic ice sheet and determining the effect on sea level using `r.lake`.

12.6   Clipping Features

Sometimes you want to create or modify a dataset by constraining it to an area of interest. We call this *clipping*, and you can do it for both raster and vector datasets using GRASS. In this section, we'll look at clipping the "collars" from a USGS DRG to allow them to display nicely side by side. We'll also look at clipping vector features to create a subset of a larger dataset.

You can also clip using the GDAL utilities. See the Linfiniti blog post at `http://goo.gl/rs5Nk` for an example of clipping a raster

Clipping Rasters with GRASS

If you are wondering why would we want to clip rasters in the first place, let me give you an example. When you download a DRG from the USGS or other source, more than likely it will have collars around the image. A collar is that nice white paper border (well, it would be paper if it wasn't digital) that contains information about the map, including the quadrangle, scale, date published, and other tidbits of information. This is all good information, except when we want to display more than one of these rasters side by side. In that case, we end up with the situation shown in Figure 12.6, on the next page. The collar of the DRG on top of the map stack blots out information from the DRG below it. To make a seamless data display, we need to remove the collars.

In its original form, a DRG looks just like the paper map you could buy from your local map store. That's because the USGS DRGs are scanned from those original maps and include not only the good stuff (contours, lakes, rivers, and so forth) but also the metadata as we indicated previously. When you plop them into GRASS, QGIS, or another GIS application, they look just like you threw them down on the kitchen table and tried to match them up. Fortunately, GRASS provides a fairly easy way to make the maps play nicely with each other.

The steps to clip a raster are as follows:

1. Create a vector map to be used as the area of interest.
2. Convert the vector map to a raster map.
3. Use the new raster map as a mask for clipping.
4. Create the newly clipped raster using raster algebra.
5. Clean things up.

Let's work through the process and see whether we can't make our rasters fit together nicely. First we have to import the rasters into GRASS in a proper location. Generally you'll find your DRG is in UTM coordinates. You'll need a GRASS location in the appropriate UTM zone in order to import the raster. If you are fortunate enough

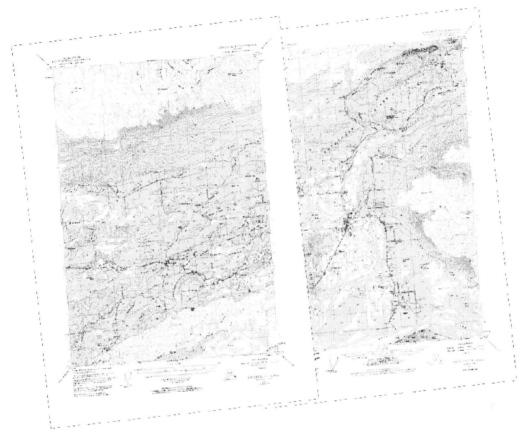

Figure 12.6: Overlapping collars on DRGs

to be working with data all in the same zone, then you're all set. If your rasters span UTM zones or you are working on a more regional scale, you may need to project the rasters to a different coordinate system. You can easily do this before you import into GRASS using `gdalwarp`. For examples of `gdalwarp`, see Section 13.2, *Using GDAL and OGR*, on page 212.

For the sake of our example, I'm going to project the DRGs to the Alaska Albers projection, since that's where I ultimately want to use them. This will eventually allow me to create a seamless DRG layer for the whole state. To warp the DRGs from UTM Zone 6 to Alaska Albers, I used the following command:

```
gdalwarp -t_srs "+proj=aea +lat_1=55 +lat_2=65 +lat_0=50 +lon_0=-154 \
  +x_0=0 +y_0=0 +ellps=clrk66 +datum=NAD27" i61149c6.tif i61149c6_albers.tif
```

Now you're thinking that doesn't look so simple, but the ugly-looking part of the command comes from the need to specify the projection in `proj` format. Since the Alaska Albers coordinate system in meters (NAD 27 datum) doesn't have an EPSG code, we have to spell it all out. If you're lucky, there will be an EPSG code for your target projection, and you can just use the EPSG:srid notation with `gdalwarp` to project the raster. For example, had I wanted to use map units of feet, the EPSG projection 2964 is perfect, and the `gdalwarp` command would have been this:

```
gdalwarp -t_srs EPSG:2964 i61149c6.tif i61149c6_albers.tif
```

The GRASS Python GUI provides all the tools needed to do the import as well as the other operations we have been discussing. If you enter the command without parameters, the GUI will popup for you.

We're now ready to import the DRG into our Alaska Albers GRASS location that we already have set up. To do this, we'll use `r.in.gdal` from the GRASS shell. In fact, this whole process will be done using shell commands rather than the GUI interface. Once we're done, we can check the results by using loading the rasters into GRASS or QGIS. To import the raster, do use:

```
r.in.gdal input=i61148e8_albers.tif output=ancc6_collars
r.in.gdal input=i61149e3_albers.tif output=ancc7_collars
```

Now we have the raster complete with collars in GRASS. The next thing we need is a vector area map that outlines just the "good" portion of the DRG in which we are interested. In most cases, you can find a vector quadrangle boundary layer somewhere on the Internet that is perfectly suited for this task. If not, you'll have to warm up your GRASS digitizing skills and create a new vector map by digitizing the four corners of the DRG. If you do find a vector layer of the quadrangles for your area, you have a bit of work to do as well, since we want only one of the quadrangle polygons. In the case of Alaska we have a quadrangle vector map named `itma` with 2,915 polygons, representing the 1:63,360 scale quadrangles. To clip the DRG, we need to create a new vector map by extracting the quadrangle of interest. We do this using the `v.extract` command:

```
v.extract input=itma output=ancc6  where="TILE_NAME='ANCC6'"
```

The input map is itma, and it contains the quadrangle boundaries. We want to create a new map named ancc6 with the boundary of the Anchorage C6 quadrangle. Notice the key part of the v.extract command—the where clause. This tells GRASS to extract only features where the attribute TILE_NAME is equal to "ANCC6," giving us a single polygon, which is what we want.

To use the boundary of the quadrangle as a mask, our new vector map has to be converted to a raster using v.to.rast:

```
g.region vect=ancc6
v.to.rast input=ancc6 output=ancc6_itma use=val
```

This creates a raster map named ancc6_itma that covers the area of the polygon in ancc6. The use=val parameter tells GRASS to set the cells to the value specified by the value parameter. Of course, you noticed that we didn't specify a value parameter. That's because it defaults to 1 if not specified, and this is exactly what we want. If we were to load up the ancc6_itma raster in GRASS or QGIS and look at the cell values, we'd find that they are indeed all set to 1. This is important—when we use this map as a mask, only those cells lying in our area of interest will have a value of 1. When we get to the final step, this will cause every cell in our DRG outside the bounds of ancc6_itma to be set to null, effectively stripping the collars.

We are now ready to actually do the clipping operation. First we set the GRASS region to that or our ancc6_collars DRG:

```
g.region rast=ancc6_collars
```

We then use the ancc6_itma raster we created from our vector quadrangle boundary as a mask:

```
g.copy ancc6_itma,MASK
```

Now that the mask is set, we use a very simple bit of map algebra to create the clipped DRG:

```
r.mapcalc ancc6=ancc6_collars
```

Notice the r.mapcalc operation looks like it just creates a new raster named ancc6 from every cell in our original DRG. The magic is in the mask, which controls which cells in the new raster are set to the same values as those in the original. Cells outside the mask are set to null. The last step is to remove the mask:

```
g.remove MASK
```

Repeating the process for the adjacent DRG (ANCC7) gives us two rasters that we can now display seamlessly, as shown in Figure 12.7. Comparing this to what we started with in Figure 12.6, on page 191, you can see that we have attained success. If we wanted to, we could combine the DRGs into a single raster using r.patch, similar to the method described in Section 12.5, *Merging Digital Elevation Models*, on page 187.

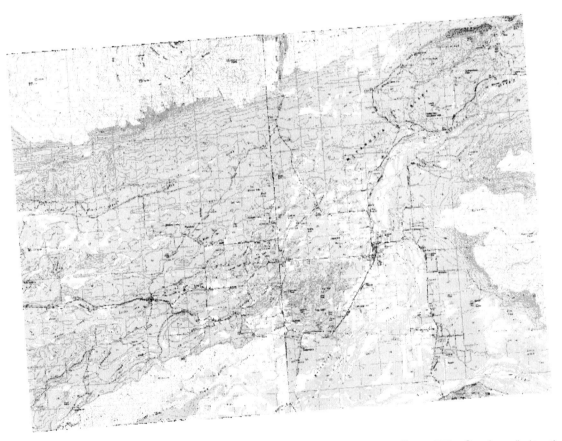

Figure 12.7: Seamless display of clipped DRGs

To put it all together, the sequence of commands we used to get
from overlapping to seamless nirvana is shown here in the form of
a bash script:

clip_raster.sh

```
# import the DRG
r.in.gdal input=i61149e3_albers.tif output=ancc7_collars
# Extract the quad boundary from the boundary map
v.extract input=itma output=ancc7 where="TILE_NAME='ANCC7'"
# Set the region to that extracted vector
g.region vect=ancc7
# Convert the extracted vector quad feature to a raster map
v.to.rast input=ancc7 output=ancc7_itma use=val
# Set the region to operate on to that of our DRG
g.region rast=ancc7_collars
# Set the mask for the operation to the raster created from the
# quad boundary vector
g.copy ancc7_itma,MASK
# Use map algebra to create the "clipped" raster
r.mapcalc ancc7=ancc7_collars
# Delete the mask
g.remove MASK
```

Clipping Vectors with GRASS

Clipping a vector map in GRASS is simpler than the raster exercise
we just went through. Basically, we need to specify the map we
want to clip and the map to be used as the clipping layer. Once we
have our data in order, we'll use the v.overlay command to do the
work.

In this example, we will clip out the rivers that are contained in a
single quadrangle. Our starting situation is shown in Figure 12.8,
on the following page, with the quadrangle shown in yellow. The
first task is to extract just the TANA5 quadrangle from our itma
quadrangle boundary map:

```
v.extract input=itma output=tana5 where="TILE_NAME='TANA5'"
```

This gives us the new vector map tana5 that contains just the quadrangle of interest. This is the same method we used in our raster
clipping process. To clip out the rivers, we simply do an intersection
of the two maps using the v.overlay command:

Figure 12.8: Rivers and the quadrangle for clipping

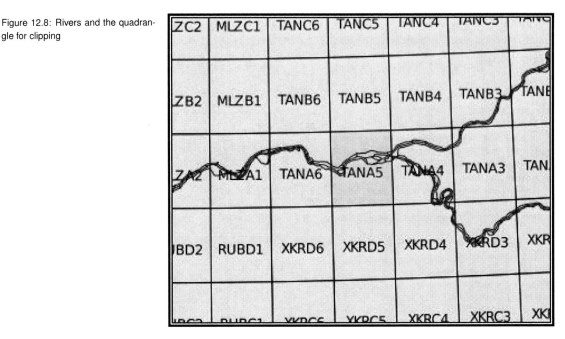

```
v.overlay ainput=majrivers atype=line binput=tana5 operator=and \
  output=majrivers_tana5
```

In Figure 12.9, on the next page, you can see the result of the clipping operation. The rivers that fall within the Tanana A5 quadrangle are all that remain in our new vector layer (majrivers_tana5). We've also included the Tanana A5 quadrangle as a backdrop in the figure. Looking at the v.overlay command, you can see that we specified the majrivers map as the first input and indicated its type using the atype parameter. The tana5 vector map we created using v.extract was specified as the second input map using the binput parameter. The key in this operation is declaring the proper operator (and) since v.overlay has four possibilities.

Remember, you can also use ogr2ogr to clip vector layers. This is especially helpful when working with non-GRASS data.

Clipping features from larger map layers to create smaller ones is a common GIS operation, especially when your project is focused in a smaller area and you don't need all the extra features running around your map. The GRASS v.overlay command provides a quick and easy way to subset your data into new map layers.

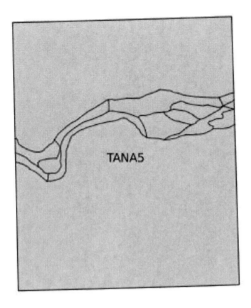

Figure 12.9: Rivers clipped to a quadrangle boundary

We'll talk about some other vector overlay operations when we get
to Chapter 14, *Getting the Most Out of QGIS and GRASS Integration*,
on page 237.

13

Using Command-Line Tools

Command-line tools provide a powerful way to manipulate data, especially when you want to process in batches using a script. This chapter describes some of the more common and useful command-line tools and illustrates how to use them to perform common data manipulation, conversion, and map generation tasks. We will take a look at the following:

- Generic mapping tools (GMT)
- Converting and appending data using GDAL/OGR
- PostGIS

13.1 GMT

For a very brief introduction to GMT, see Appendix 16, *Appendix A: Survey of Desktop GIS Software*, on page 299. In this section, we will take a look at using GMT to create nicely formatted maps for displaying and printing. But before we can do that, we need to make sure you have GMT installed. If not, take a look at Section 17.5, *GMT*, on page 320 for some hints to get you started.

The GMT commands create Encapsulated PostScript (EPS) output. If you are using Linux or OS X, you should already have the tools you need to view eps files. On Windows you will need a viewer that supports EPS. One such viewer is GSview, which allows EPS

files to be viewed and printed. For other options, use your favorite
Internet search engine to find a suitable application that works for
you.

A Simple GMT Example

Figure 13.1: Hemisphere view of
Earth created with GMT

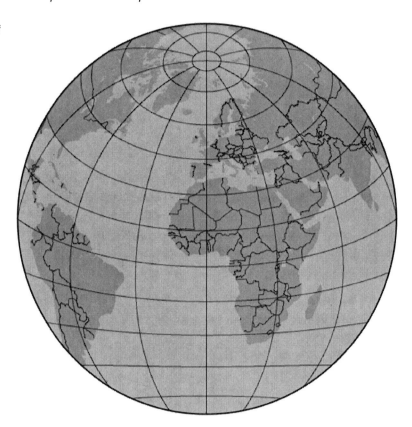

To get started, let's take a look at how to generate a simple globe
like that shown in Figure 13.1. The code is pretty simple, although
it's a bit arcane at first glance:

```
pscoast -JA0/20/4.5i -Bg30/g15 -Dl -A2000 -G187/142/46 -S109/202/255 \
  -R0/360/-90/90 -P -N1 > simple_hemi.eps
```

Let's examine the switches used to generate the image. First, GMT's
pscoast command requires information about the coordinate system you want to use. This is specified by using the -J switch. GMT

supports a nice selection of coordinate systems including the following:

- Albers Conic Equal Area
- Lambert Conic Conformal
- Equidistant Conic
- Lambert Azimuthal Equal Area
- Stereographic Equal Angle
- Orthographic
- Azimuthal Equidistant
- Gnomonic
- Mercator
- Transverse Mercator
- Universal Transverse Mercator
- Oblique Mercator
- Cassini Cylindrical
- Cylindrical Equidistant
- General Cylindrical
- Miller Cylindrical
- Miscellaneous

Each projection has a specific argument that must be supplied to the -J switch. Looking back at the globe example, you'll see that -JA was used to specify the Lambert Azimuthal Equal Area projection. I know because the programs that make up GMT provide a complete description of what's expected as input when you run them with no options. For example, if we enter the pscoast command, we get several screens of options and switches. The first part contains the available projection switches and their syntax:

```
$ pscoast
pscoast 4.5.7 [64-bit] - Plot continents, shorelines, rivers, and borders on maps

usage: pscoast -B<params> -J<params>
[-A<min_area>[/<min_level>/<max_level>][+r|l][+p<percent>]]
[-R<west>/<east>/<south>/<north>[/<zmin/zmax>][r]]
    [-C[<feature>/]<fill>]
[-D<resolution>][+]
[-E<azim>/<elev>[+w<lon>/<lat>[<z>][+v<x0>/<y0>]]
[-G[<fill>]]
```

```
[-I<feature>[/<pen>]]
[-Jz|Z<params>]
[-K]
[-L[f][x]<lon0>/<lat0>[/<slon>]/<slat>/<length>[m|n|k][+l<label>][+j<just>]
    [+p<pen>][+f<fill>][+u]]
[-N<feature>[/<pen>]]
[-O] [-P] [-Q]
[-S<fill>]
[-T[f|m][x]<lon0>/<lat0>/<diameter>[/<info>][:w,e,s,n:][+<gint>[/<mint>]]]
[-U[<just>/<dx>/<dy>/][c|<label>]]
[-V]
[-W[<feature>/][<pen>]]
[-X[a|c|r]<x_shift>[u]]
[-Y[a|c|r]<x_shift>[u]]
[-Z<zlevel>]
[-bo[s|S|d|D[<ncol>]|c[<var1>/...]]]
[-c<ncopies>]
[-m[<flag>]]
-J Selects map proJection. (<scale> in cm/degree, <width> in cm)
    Append h for map height, + for max map dimension, and - for min map
    dimension.  Azimuthal projections set -Rg unless polar aspect or
    -R<...>r is given.

    -Ja|A<lon0>/<lat0>[/<horizon>]/<scale (or radius/lat)|width>
        (Lambert Azimuthal Equal Area)
   -Jb|B<lon0>/<lat0>/<lat1>/<lat2>/<scale|width> (Albers Equal-Area Conic)
    -Jcyl_stere|Cyl_stere/[<lon0>/[<lat0>/]]<lat1>/<lat2>/<scale|width>
        (Cylindrical Stereographic)
   -Jc|C<lon0>/<lat0><scale|width> (Cassini)
   -Jd|D<lon0>/<lat0>/<lat1>/<lat2>/<scale|width> (Equidistant Conic)
    -Je|E<lon0>/<lat0>[/<horizon>]/<scale (or radius/lat)|width>
        (Azimuthal Equidistant)
   -Jf|F<lon0>/<lat0>[/<horizon>]/<scale (or radius/lat)|width>  (Gnomonic)
   -Jg|G<lon0>/<lat0>/<scale (or radius/lat)|width>   (Orthographic)
    -Jg|G[<lon0>/]<lat0>[/<horizon>|/<altitude>/<azimuth>/<tilt>/<twist>/
        <Width>/<Height>]/<scale|width> (General Perspective)
   -Jh|H[<lon0>/]<scale|width> (Hammer-Aitoff)
   -Ji|I[<lon0>/]<scale|width> (Sinusoidal)
   -Jj|J[<lon0>/]<scale|width> (Miller)
   -Jkf|Kf[<lon0>/]<scale|width> (Eckert IV)
   -Jks|Ks[<lon0>/]<scale|width> (Eckert VI)
   -Jl|L<lon0>/<lat0>/<lat1>/<lat2>/<scale|width> (Lambert Conformal Conic)
   -Jm|M[<lon0>/[<lat0>/]]<scale|width> (Mercator).
   -Jn|N[<lon0>/]<scale|width> (Robinson projection)
   -Jo|O (Oblique Mercator).  Specify one of three definitions:
      -Jo|O[a]<lon0>/<lat0>/<azimuth>/<scale|width>
      -Jo|O[b]<lon0>/<lat0>/<lon1>/<lat1>/<scale|width>
      -Jo|Oc<lon0>/<lat0>/<lonp>/<latp>/<scale|width>
   -Jpoly|Poly/[<lon0>/[<lat0>/]]<scale|width> ((American) Polyconic)
   -Jq|Q<lon0>/[<lat0>/]]<scale|width> (Equidistant Cylindrical)
```

```
-Jr|R[<lon0>/]<scale|width> (Winkel Tripel)
  -Js|S<lon0>/<lat0>/[<horizon>/]<scale (or slat/scale or
     radius/lat)|width> (Stereographic)
-Jt|T<lon0>/[<lat0>/]<scale|width> (Transverse Mercator).
-Ju|U<zone>/<scale|width> (UTM)
-Jv|V<lon0>/<scale|width> (van der Grinten)
-Jw|W<lon0>/<scale|width> (Mollweide)
-Jy|Y[<lon0>/[<lat0>/]]<scale|width> (Cylindrical Equal-area)
-Jp|P[a]<scale|width>[/<origin>][r|z] (Polar [azimuth] (theta,radius))
  -Jx|X<x-scale|width>[d|l|p<power>|t|T][/<y-scale|height>[d|l|p<power>|t|T]]
     (Linear, log, and power projections)
(See psbasemap for more details on projection syntax)
```

Each projection requires different parameters. In our example we used -JA0/20/4.5i. This selects the projection (Lambert Azimuthal Equal Area) and sets the longitude to zero degrees, the latitude to 20 degrees, and the width of the map to 4.5 inches. Widths can be specified using c, i, p, or m, which correspond to centimeters, inches, points (1/72 of an inch), and meters.

You can quickly see two things:

- GMT has a lot of options.
- You might want to read the manual.

This is an important point—sometimes it takes a bit of digging and looking at examples to find the switches, arguments, or parameters needed to accomplish your goal. Reading the manual is a good place to start.

Let's see what happens if we modify the -J switch a bit. Let's flip the view around 180 degrees and move it closer to the North Pole. To do this, use -JA180/65/4.5i. Leave all the other parameters the same, and run the pscoast command. Our command is now as follows:

```
pscoast -JA180/65/4.5i -Bg30/g15 -Dl -A2000 -G187/142/46 -S109/202/255 \
  -R0/360/-90/90 -P -N1 > simple_180_world.eps
```

Looking at Figure 13.2, on the following page, you see that indeed we are now looking at the International Date Line, and our view is centered at 65 degrees north latitude.

Figure 13.2:   Globe centered on 180/65

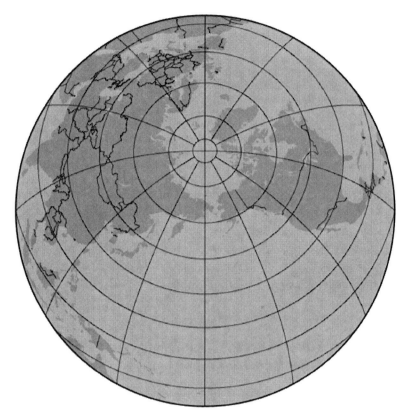

Let's take a look at the other switches used to create the globe. The -B switch defines the intervals for the boundary tick marks. In the globe case, these are the lines of longitude and latitude. The arguments to the -B switch indicate a grid line spacing of 30 degrees in the x (longitude) direction and 15 degrees in the y (latitude) direction. Note how the x and y settings are separated by a forward slash.

The -D switch selects the resolution of the dataset used in creating the globe. The available choices are f, h, i, l, and c, which correspond to full, high, intermediate, low, and crude. Some of these options may not be available to you if you didn't install all the data sets with GMT. For the globe, we used the low resolution data set.

Where Does the GMT Data Come From?

GMT actually provides a number of datasets in five resolutions ranging from fine to crude. When installing GMT, you can choose which resolutions you want to include.

There are also tools available to convert other formats for use with GMT. Use your favorite search engine to find tools applicable to your situation.

To control the display of features, the -A switch allows you to specify that features below a certain size not be drawn. In our example, we specified that features with an area greater smaller 2,000 square kilometers should not be displayed.

The fill color used for the countries is specified using the -G switch. The color can be specified using RGB notation, a shade of gray, or a pattern. In the globe, we used 187/142/46 to create a light brown color. We could have specified a fill pattern using -Gp100/30. This fills the land masses with pattern number 30 at a resolution of 100 dpi. If we want to get the highest possible resolution for the pattern, we can use a resolution of zero. Specifying -GP inverts the pattern. GMT has 90 predefined patterns available for your use, and you can find examples of each in the GMT Technical Reference. The same options apply for filling the water areas in GMT, except we use the -S switch. There are a number of variations for specifying fill colors, and these are well documented in the GMT manuals and tutorial.

The other major switch used in generating the globe is -R. This specifies the extent of the map we want to generate. In the case of the globe, we obviously wanted the entire planet, so we specified an x range of 0 to 360 degrees and a y range of -90 to 90. The range is specified as west/east/south/north. In our next example, we will use -R to constrain our map to a smaller area.

The other switch of interest is -N1. This tells GMT to draw national boundaries in addition to the coastline. Other arguments to -N allow you to draw state boundaries within the Americas and marine boundaries.

The -P switch simply sets the page orientation to portrait. Land-scape is the default.

A Flat Example

Let's shift gears a bit and look at another example of using GMT, this time for a smaller area. For this example, we'll create a map of Alaska and annotate it. As I said before, the -R switch controls the extent of our map. Alaska ranges from about 172 degrees east longitude to 130 degrees west. Using 360 degrees for the entire globe, this translates to a region extending from 172 degrees to 230 degrees.

For the Alaska map, we will use the Albers Equal Area Conic projection. Looking at the syntax for pscoast reveals that this requires the use of the -Jb switch. In this case, we use the lowercase *b* to indicate that we will specify the size of the map using a scale. First let's look at the code in gmt_alaska.sh:

```
pscoast -Jb-154/50/55/65/1:12000000 -R172/230/51/72 -B10g5/5g5 -W1p/0/0/0 \
  -I1/2p/0/192/255 -I2/2p/0/192/255 -I3/1p/0/192/255 -I4/1p/0/192/255 \
  -G220/220/220 -S0/192/255 -L210/54/54/1000 -P -N1/1p/0/0/0 -Dl \
  >gmt_alaska_coast.eps
```

This looks like quite a complex command, but it's really not too bad once you get past all the numbers and slashes.

Projection

First note we specified the projection using -Jb-154/50/55/65/1:12000000. Let's pick that apart a bit to see what's happening. The Albers projection requires the longitude of the central meridian, the latitude of the origin, and the latitude of the two standard parallels. That's what you see specified as -154/50/55/65. These are the standard values used for the Albers projection in Alaska. You can actually specify any values you want, but if there is a standard for the area you are mapping, you should use it.

The remaining part of the -Jb switch is the size of the output. In this case, we specified it as a scale of 1:12,000,000. This means that one unit on the map represents 12,000,000 units on the ground (in this case meters). If we just wanted output to fit on a page, we could specify -JB-154/50/55/65/6.0i to get a 6-inch-wide image.

If you find your map doesn't fit on the "paper" you can change the media size using gmtset. See the gmtdefaults documentation for available sizes.

Map Extent

To set the map extent, we use the -R switch. In this case we already determined that Alaska ranges from 172 to 230 degrees longitude and roughly 51 to 72 degrees north latitude. To create the map covering this area, we use -R172/230/51/72.

Grid Lines

In this example, we not only want to draw grid lines but also want to annotate them. This is done using -B10g5/5g5. This tells pscoast to draw grid lines 5 degrees apart for both latitude and longitude. The annotation is drawn at 10 degree intervals for longitude and 5 degree intervals for latitude. If you look at the documentation for pscoast, you will see that the first number after the -B is the annotation interval followed by the grid line interval. This notation gives you a lot of flexibility in drawing and labeling grid lines.

Rivers

To make our map more interesting, we'll add rivers to it. GMT comes with several levels of river detail that are specified with the -I switch. The levels we are using are as follows:

- Permanent major rivers
- Additional major rivers
- Additional rivers
- Minor rivers

Notice the -I options we specified in the pscoast command. One is required for each river level we want to include on the map. The first two (major rivers) are drawn using a pen width of 2 (2p),

208 CHAPTER 13. USING COMMAND-LINE TOOLS

while the third and fourth level are drawn with a width of 1 (1p). We use the same color (0/192/255) for each river. If we wanted to include the intermittent major rivers (fifth level), we would add -I5/1p/0/192/255 to the pscoast command.

Fill Colors

Next we specify the fill colors for the land and water areas using the -G and -S switches and add the RGB values to specify the color. For land we use a light gray with RGB values of 220/220/220. For the water 0/192/255 gives us a nice cyan color. Keep in mind that we could also use a pattern or shade for filling land and water areas.

Scale Bar

A scale bar can easily be added to the map using the -L switch. Scale bars can be simple or fancy. In this case, we'll just create a simple one and place it in an open area on the map. How do we know it's open? Well, part of the process is running pscoast and tweaking the options and then running it again until we get the look we want. To create the scale bar, we need the latitude and longitude of the point where we want to place it. Since scale varies as we move further from the equator, we also specify the latitude at which we want the scale calculated. Lastly, we indicate the length the scale bar should span. The default is kilometers, but you can append m for miles or n for nautical miles. Putting it all together, we have -L210/54/54/1000, which gives a 1,000 km scale bar calculated at 54 degrees north latitude and originating at 210 degrees longitude and 54 degrees latitude.

The Last Bits

The remainder of the command tells pscoast to use portrait mode (-P), draw country boundaries in black using a pen width of 1 (-N1/1p/0/0/0), and use the low-resolution data (-Dl). The low-resolution dataset is the default, but we specified it here so you could see the syntax.

The Result

You can see the result of all these command switches and options in Figure 13.3. We have a nice map of Alaska, with grid lines, borders, and degree annotations. The land and water is filled as we specified, and the scale bar is sitting nicely in the Gulf of Alaska.

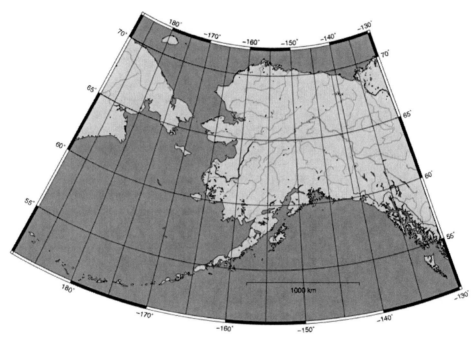

Figure 13.3: Alaska coastline gener-ated with GMT

Overlaying Data

Now that we have used most of the common options, let's look at one more example with pscoast. This time we'll generate a world map and overlay point data from a delimited text file. You can take that concept and expand it to create overlays of multiple datasets. In this case, we will overlay the location of the world's volcanoes. First let's look at the command to generate the base map:

```
pscoast -JN0/38 -R-180/180/-90/90 -W -G220/220/220 -S0/192/255 -N1 \
    -P -B30g5:."World Volcanoes": > world_volcanoes.eps
```

About the only thing new in this command is we added a title to the map by appending a colon and a period to the -B arguments and then the title string. If you are getting the idea the -B switch has lots of permutations, you are correct. Some have called it the most complicated (or confusing) switch in the GMT suite of tools. Fortunately, it's well documented.

Note that we used -JN to specify the Robinson projection, centering the map at 0 degrees longitude with a width of 38 centimeters.

GMT and Multiple Commands

When you want to create a more complex map, you will wind up running multiple GMT commands. The trick is to make sure you specify in the first command that more PostScript code will be appended to the output. Without this, you will end up with every command generating a new page in the output. This can be useful, but when you are trying to create an overlay of multiple commands, it's annoying. Create the base map using the -K switch, and then in subsequent commands include the -0 to invoke overlay mode.

This gives us a Robinson base map of the world with grid lines and annotation of the tick marks. To add an overlay of data, we need to add a couple of things. First we need to specify that we want to be able to write to the output file in "append" mode. This is done using the -K switch. This allows us to overlay the data created with the next command. Without it, our next command would generate a new page in the output, and we would have to hold it up to the window to see the overlay.

Secondly we need to use the psxy command with a text file containing the coordinates of each volcano. Here is a snippet from the text file we will use to create the map:

```
131.6,34.5
-67.62,-23.3
-90.876,14.501
34.52,38.57
-69.05,-18.37
```

-176.6,51.98

These are just x and y values (longitude and latitude) for each point we want to create on the map. To add these, we use psxy, making sure to include the -O switch so the points overlay the base map. The complete code to generate the base map and the overlay is as follows:

```
pscoast -JN0/38 -R-180/180/-90/90 -K -W -G220/220/220 -S0/192/255 -N1 \
    -P -B30g5:."World Volcanoes": > world_volcanoes.eps
psxy volcanoes_lon_lat.txt -JN -O -R -Sc0.15c -G255/0/0 >> world_volcanoes.eps
```

Note that in the psxy command we didn't need to supply any arguments to the projection or extent switches since they were fully specified when the base map was created. We included the overlay switch and a color for the points using -G. The -Sc switch indicates we want to plot the points using circles. There are several other symbol types you can use including star, bar, diamond, and ellipse. We specified the symbol size as 0.15 cm. You should recognize the rest of the parameters from our previous discussion.

The final result is shown in Figure 13.4, on the following page. This combines the base map with the volcano data. We could have added other point data by running another psxy command.

As you can see, GMT is a handy tool for creating maps from the command line. We've really only scratched the surface of its capability.

More GMT

GMT has many more features than we have covered—there are over 100 commands. You should have the basics down, allowing you to venture forward and generate even more impressive maps. Make sure to consult the GMT documentation for additional information, including cookbook recipes and tutorials.

For additional information on using GMT with GRASS, see "Producing Press-Ready Maps with GRASS and GMT" by Dylan Beaudette in *OSGeo Journal*, Volume 1, May 2007.[1]

[1] http://www.osgeo.org/files/journal/final_pdfs/OSGeo_vol1_GRASS-GMT.pdf

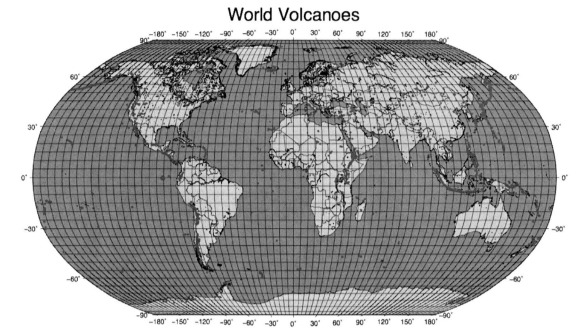

Figure 13.4: Volcanoes plotted on a
Robinson projection using GMT

13.2   Using GDAL and OGR

We have seen examples of the GDAL and OGR utilities previously
in several sections. Now we will take a more focused look at the
utilities and how they are used. You will quickly see that this set of
tools belongs in your toolkit, especially if you plan to do any data
manipulation.

If you want a quick overview of the formats supported by GDAL
and OGR, as well as a brief summary of each utility, see Section 16.2,
GDAL/OGR, on page 309 in our survey of OSGIS software.

Getting Information

One of the key uses of the GDAL/OGR utilities is getting information about a supported vector or raster file. The commands used
are ogrinfo and gdalinfo, respectively. Let's take a better look at
each of these utilities.

Vector Information

You download a shapefile from the Internet and unzip it. Now you have a batch of files sitting there (remember, a shapefile consists of at least three files). What attributes does the shapefile contain? What kind of features does it store—points, lines, or polygons? What projection or coordinate system does it use? We are in luck; ogrinfo can answer all those questions for us.

The ogrinfo utility can provide information on both a single layer and all layers in a directory. For example, for a summary of all shapefiles in a directory, we just have to provide the directory name:

```
$ ogrinfo ./desktop_gis_data
INFO: Open of `./desktop_gis_data'
using driver `ESRI Shapefile' successful.
1: cities (Point)
2: AKvolc_v3 (Point)
3: world_borders (Polygon)
```

From the output we can see that the directory desktop_gis_data contains three shapefiles: cities, AKvolc_v3, and world_borders. The results for each shapefile includes its type. This is good summary information, but what if we want more detail? By specifying the layer name, ogrinfo will give us very detailed information about the layer:

```
$ ogrinfo -so -al ./desktop_gis_data cities
INFO: Open of `./desktop_gis_data'
using driver `ESRI Shapefile' successful.

Layer name: cities
Geometry: Point
Feature Count: 606
Extent: (-165.270004, -53.150002) - (177.130188, 78.199997)
Layer SRS WKT:
GEOGCS["GCS_WGS_1984",
    DATUM["WGS_1984",
        SPHEROID["WGS_1984",6378137,298.257223563]],
    PRIMEM["Greenwich",0],
    UNIT["Degree",0.0174532925199433]]
NAME: String (40.0)
COUNTRY: String (12.0)
POPULATION: Real (11.0)
CAPITAL: String (1.0)
```

The result gives us a detailed summary of information about the cities layer. We can see it is a point layer with 606 features. The coordinate system is WGS84, meaning the coordinates are in latitude and longitude. We also get a summary of the fields and their types, along with the extent of the layer. Armed with this information, we can easily determine whether a layer is suitable for our use and is in an appropriate coordinate system.

Notice the -so switch in the previous example. We used it in combination with the -al switch in order to get detailed information about the layer. The -so switch tells ogrinfo to print a summary only; otherwise, it would also print each record in the shapefile, complete with all the attributes as well as the coordinate information. There are times you may want to view all the information, perhaps dumping it to a text file for further use.

We can use the OGR utilities with more than just shapefiles. To get a quick list of the supported drivers for your installation of OGR, use the --formats switch. The formats you find available will depend on how your version of OGR was compiled. If our version contains support for PostgreSQL, we can get information on layers stored in our PostGIS-enabled database:

```
$ ogrinfo "PG:dbname=gis_data host=madison"
INFO: Open of `PG:dbname=gis_data host=madison'
using driver `PostgreSQL' successful.
1: edit_test (Point)
2: country (Multi Polygon)
3: air_intl_buffer_500k12 (Polygon)
4: bug_test (Polygon)
5: air_intl_buffer_500k14 (Polygon)
6: 64districts (Multi Polygon)
7: 94election (Multi Polygon)
8: admin_nps (Multi Polygon)
9: admin_nra (Multi Polygon)
10: admin_nwr (Multi Polygon)
...
```

An important part of using ogrinfo with PostGIS is the connection string, specified with "PG:" and followed by the appropriate parameters. In our case, we needed only to specify the database name and the host. Depending on how your database authentication is set up, you may need to include "user=" and "password=" (with the appropriate values) in your connection string.

This database contains more than 100 layers, so we truncated the listing—but you get the idea. We didn't have to connect to the database, log in, and issue some SQL to determine what layers were available. We didn't even have to be on the same host as the database in order to get the information—of course this assumes you have a properly set up database with appropriate permissions. Let's get the details for the `country` layer, remembering to use the `-so` switch so we don't dump the whole world to our command shell:

```
$ ogrinfo -so -al "PG:dbname=gis_data host=madison" country
INFO: Open of `PG:dbname=gis_data host=madison'
using driver `PostgreSQL' successful.

Layer name: country
Geometry: Multi Polygon
Feature Count: 251
Extent: (-180.000000, -90.000000) - (180.000000, 83.623596)
Layer SRS WKT:
GEOGCS["WGS 84",
    DATUM["WGS_1984",
        SPHEROID["WGS 84",6378137,298.257223563,
            AUTHORITY["EPSG","7030"]],
        AUTHORITY["EPSG","6326"]],
    PRIMEM["Greenwich",0,
        AUTHORITY["EPSG","8901"]],
    UNIT["degree",0.01745329251994328,
        AUTHORITY["EPSG","9122"]],
    AUTHORITY["EPSG","4326"]]
Geometry Column = shape
cntry_name: String (40.0)
color_map: String (1.0)
curr_code: String (4.0)
curr_type: String (16.0)
fips_cntry: String (2.0)
gid: Integer (0.0)
gmi_cntry: String (3.0)
landlocked: String (1.0)
pop_cntry: Integer (0.0)
sovereign: String (40.0)
```

The output looks similar to that for a shapefile. The thing to note is, in addition to the fields in the layer, `ogrinfo` also identifies the name of the geometry column for us—in this case it's shape.

Raster Information

For getting information on your rasters, gdalinfo is the tool to use. We introduced this back in Section 6.1, *Viewing Raster Data*, on page 73 where we examined a GeoTIFF. Let's look at some of the options and formats associated with the utility.

To get a list of all the supported formats at your disposal, use the --formats switch. When you do this, you're likely going to get a long list. I won't list all 104 found on my system, but just the first few as an example:

```
GRASS (ro): GRASS Database Rasters (5.7+)
  VRT (rw+v): Virtual Raster
  GTiff (rw+v): GeoTIFF
  NITF (rw+v): National Imagery Transmission Format
  RPFTOC (rov): Raster Product Format TOC format
  HFA (rw+v): Erdas Imagine Images (.img)
  SAR_CEOS (rov): CEOS SAR Image
  CEOS (rov): CEOS Image
  JAXAPALSAR (ro): JAXA PALSAR Product Reader (Level 1.1/1.5)
  GFF (rov): Ground-based SAR Applications Testbed File Format (.gff)
  ELAS (rw+): ELAS
  AIG (rov): Arc/Info Binary Grid
  AAIGrid (rwv): Arc/Info ASCII Grid
  SDTS (rov): SDTS Raster
  DTED (rwv): DTED Elevation Raster
  PNG (rwv): Portable Network Graphics
  JPEG (rwv): JPEG JFIF
```

We can use gdalinfo on files that aren't "strictly" GIS files. For example, here is the output for a JPEG from a digital photo:

```
gdalinfo -mm DSCN3898.jpg
Driver: JPEG/JPEG JFIF
Files: /Users/gsherman/DSCN3898.jpg
Size is 685, 1024
Coordinate System is ''
Metadata:
  EXIF_PhotometricInterpretation=32803
  EXIF_Make=NIKON CORPORATION
  EXIF_Model=NIKON D3000
  EXIF_Orientation=1
```

```
  EXIF_XResolution=(72)
  EXIF_YResolution=(72)
  EXIF_ResolutionUnit=2
  EXIF_Software=Aperture 3.1.3
  EXIF_DateTime=2011:10:09 13:52:21
  ...
Image Structure Metadata:
  SOURCE_COLOR_SPACE=YCbCr
  INTERLEAVE=PIXEL
  COMPRESSION=JPEG
Corner Coordinates:
Upper Left  (    0.0,    0.0)
Lower Left  (    0.0, 1024.0)
Upper Right ( 685.0,    0.0)
Lower Right ( 685.0, 1024.0)
Center      ( 342.5,  512.0)
Band 1 Block=685x1 Type=Byte, ColorInterp=Red
    Computed Min/Max=0.000,255.000
  Image Structure Metadata:
    COMPRESSION=JPEG
Band 2 Block=685x1 Type=Byte, ColorInterp=Green
    Computed Min/Max=15.000,255.000
  Image Structure Metadata:
    COMPRESSION=JPEG
Band 3 Block=685x1 Type=Byte, ColorInterp=Blue
    Computed Min/Max=22.000,255.000
  Image Structure Metadata:
    COMPRESSION=JPEG
```

There was a lot more metadata in the output, sixty-eight lines in total. Basically, every EXIF field the camera stored was dumped. The point is, gdalinfo can provide detailed information for the formats it supports—and, yes, just in case you were wondering, you can georeference a JPEG. You won't see any coordinate system information for this digital photo; however, if there was a world file associated with it, the information would be included in the output. For cameras with a GPS, the latitude and longitude are encoded in the EXIF data and gdalinfo will display the coordinates.

Using gdalinfo is a quick and efficient way to get information on your rasters, without having to open them in your GIS application. For complete information on all the options, see the GDAL documentation.[2]

[2] http://www.gdal.org/gdalinfo.html

Converting Data

The GDAL and OGR utilities allow you to convert data between formats, optionally changing some of the characteristics in the process. In this section, we'll look at options and techniques for data conversion, both raster and vector.

Vector Conversion

First off, let's think about why you might want to convert from one vector format to another:

- You have data in a format that isn't usable in your desktop GIS application.
- You need to provide data (to someone or to another application) in a different format than what you are using.
- The data is in the wrong coordinate system or datum.
- You want to create a subset of the data based on a bounding box.
- You want to create a subset of the data based on an attribute query.
- You want to load data into PostgreSQL/PostGIS.

These all sound like good reasons for doing a conversion. Let's look at a few simple examples to get you started with ogr2ogr.

Format Conversion   First let's convert a vector layer from one format to another. In the simplest case, we specify the format we want to convert to, the source layer, and the destination:

```
ogr2ogr -f GML cities.gml cities.shp
```

[3] Geographic Markup Language—see http://www.opengeospatial.org/standards/gml.

We just converted the cities.shp shapefile to GML.[3] Notice there was no output or confirmation from ogr2ogr, but the file was created and contains all features in the cities layer in GML. Going the other direction is just as easy:

```
ogr2ogr -f "ESRI Shapefile" cities_from_gml.shp cities.gml
```

Notice that we specify the format we are converting *to* using the -f switch. Remember you can get a list of supported formats passing the - -formats switch to ogr2ogr. If the format name contains spaces, you'll have to quote it as we did for the ESRI shapefile conversion.

Data Loading We can use ogr2ogr to load data into a PostGIS-enabled PostgreSQL database. If you are loading just shapefiles, you could use the shp2pgsql utility that comes with PostGIS. Otherwise, you will find ogr2ogr handy for loading other data types. Let's load the GML file we created into PostgreSQL:

In the following examples we will be using two different databases. One is named gis_data located on the server "madison" and the other is the geospatial_desktop database on the local host.

```
ogr2ogr -f PostgreSQL -a_srs EPSG:4326 "PG:dbname=gis_data host=madison" \
  cities.gml
ogrinfo -so -al "PG:dbname=gis_data host=madison" cities
INFO: Open of 'PG:dbname=geospatial_desktop'
      using driver 'PostgreSQL' successful.

Layer name: cities
Geometry: Point
Feature Count: 606
Extent: (-165.270004, -53.150002) - (177.130188, 78.199997)
Layer SRS WKT:
GEOGCS["WGS 84",
    DATUM["WGS_1984",
        SPHEROID["WGS 84",6378137,298.257223563,
            AUTHORITY["EPSG","7030"]],
        AUTHORITY["EPSG","6326"]],
    PRIMEM["Greenwich",0,
        AUTHORITY["EPSG","8901"]],
    UNIT["degree",0.01745329251994328,
        AUTHORITY["EPSG","9122"]],
    AUTHORITY["EPSG","4326"]]
FID Column = ogc_fid
Geometry Column = wkb_geometry
capital: String (1.0)
name: String (40.0)
country: String (12.0)
population: Real (0.0)
```

Well, that worked—the GML file was successfully loaded into PostgreSQL. Using ogrinfo, we confirmed that it was loaded and had the proper coordinate system. If you look at the load command, you will notice we specified the coordinate system with the -a_srs switch using an EPSG code of 4326 (WGS 84, latitude/longitude). We did this because the GML file contains no projection information, even though the shapefile we created it from did. This allows our newly created PostGIS layer to play nicely with other WGS 84 layers in our database. Note that by specifying -a_srs, we aren't transforming or changing the data in any way. All we are doing is assigning the coordinate system to the PostGIS layer when it is created.

As you suspected, we can also unload data from PostGIS to a supported format. For example, we'll unload one of our layers to GML:

```
$ ogr2ogr -f GML volcanoes.gml "PG:dbname=gis_data host=madison" volcanoes
```

We can also specify a where clause in ogr2ogr to create a subset of our data. This works quite well for extracting features from PostGIS where the database might be quite large and we need only a small set of the data for our purpose. To extract a subset, use the -where switch and enclose the clause in double quotes:

You'll notice here we are specifying the schema to use for the volcano layer with the active_schema parameter.

```
$ ogr2ogr -f "ESRI Shapefile" strato_volcanoes.shp \
    "PG:dbname=geospatial_desktop active_schema=hazards" \
    volcanoes -where "type = 'Stratovolcano'"

$ ogrinfo -so -al strato_volcanoes.shp
INFO: Open of 'strato_volcanoes.shp'
      using driver 'ESRI Shapefile' successful.

Layer name: strato_volcanoes
Geometry: Point
Feature Count: 713
Extent: (-178.800000, -77.530000) - (179.620000, 71.080000)
Layer SRS WKT:
GEOGCS["GCS_WGS_1984",
    DATUM["WGS_1984",
        SPHEROID["WGS_1984",6378137,298.257223563]],
    PRIMEM["Greenwich",0],
    UNIT["Degree",0.017453292519943295]]
longitude: Real (24.15)
elev: Integer (10.0)
country: String (80.0)
type: String (80.0)
name: String (80.0)
region: String (80.0)
latitude: Real (24.15)
status: String (80.0)
last_erupt: String (80.0)
number: String (80.0)
```

We just created a shapefile containing only volcanoes of type Stra-
tovolcano from our original layer in PostGIS. The PostGIS layer contains 1,545 volcanoes and our use of the where clause whittled that
down to 713 matches. If all you want to do is display a subset of
a PostGIS layer, remember that QGIS supports subsets on the fly;
otherwise, this is another useful technique for moving data around.

Another option when unloading or converting data is to specify a
bounding rectangle using the -spat option. This will allow you to
extract only those features within the rectangle, creating a spatial
subset. You need to specify the bounds of the rectangle in the same
coordinate system as the layer. To illustrate, let's extract a small
subset of the cities layer we loaded into PostGIS and list the results
using ogrinfo.

```
$ ogr2ogr -f "ESRI Shapefile" cities_subset.shp \
  -spat -152 58 -148 62 "PG:dbname=gis_data host=madison" cities
$ ogrinfo  -al cities_subset.shp
INFO: Open of `cities_subset.shp'
      using driver `ESRI Shapefile' successful.

Layer name: cities_subset
Geometry: Point
Feature Count: 2
Extent: (-149.449997, 60.119999) - (-149.172974, 61.188648)
Layer SRS WKT:
GEOGCS["GCS_WGS_1984",
    DATUM["WGS_1984",
        SPHEROID["WGS_1984",6378137,298.257223563]],
    PRIMEM["Greenwich",0],
    UNIT["Degree",0.017453292519943295]]
capital: String (1.0)
country: String (12.0)
name: String (25.0)
population: Real (11.0)
OGRFeature(cities_subset):0
  capital (String) = N
  country (String) = US
  name (String) = Seward
  population (Real) =        2699
  POINT (-149.449996948241989 60.119998931884801)

OGRFeature(cities_subset):1
  capital (String) = N
  country (String) = US
  name (String) = Anchorage
```

```
population (Real) =    184300
POINT (-149.172973632811988 61.188648223877003)
```

The `cities` layer in PostGIS has 606 features. You can see that our spatial subset has two features, both contained within the latitude/-longitude rectangle we specified. Notice that the extent of the new layer is less than what we specified as the spatial boundaries. The bounding rectangle is specified as xmin, ymin to xmax, ymax—in this case -152, 58 to -148, 62. The extent of the new layer is smaller, because it represents the extent of the features *extracted*, not the search rectangle. You can probably think of ways in which creating subsets by attributes or spatial boundaries can come in handy.

The OGR utilities support KML which means you can export an OGR supported data source for use in Google Earth.[4]

```
$ ogr2ogr -f KML volcanoes.kml "PG:dbname=gis_data host=madison" volcanoes
```

Coordinate System Conversion   We can also change the coordinate system of a layer using `ogr2ogr`. You can do this even if you don't want to change the format of the layer. For our `cities.shp` we created earlier, we could convert it from WGS 84 (latitude/longitude) to some other coordinate system, such as U.S. National Atlas Equal Area. To do this, we need to know either the projection parameters or the EPSG code. As we saw in Chapter 11, *Projections and Coordinate Systems*, on page 159, there are a number of ways to find this—additional ways include querying the `spatial_ref_sys` table in PostGIS and using the projection search feature in the QGIS projection dialog box. If you like, you can download all the EPSG codes in several database formats from OGP.[5] Another handy reference for coordinate systems is the Spatial Reference website, which provides an interactive web interface that allows you to find and display spatial reference information.[6]

Let's convert the cities shapefile to the U.S. National Atlas Equal Area projection (EPSG:2163) and check the result:

```
ogr2ogr -t_srs EPSG:2163 cities_2163.shp cities.shp
```

```
ogrinfo -so -al cities_2163.shp
```

```
INFO: Open of 'cities_2163.shp'
      using driver 'ESRI Shapefile' successful.

Layer name: cities_2163
Geometry: Point
Feature Count: 606
Extent: (-11998953.492272, -9403336.326458) - (11911509.773028, 11451428.745740)
Layer SRS WKT:
PROJCS["Lambert_Azimuthal_Equal_Area",
    GEOGCS["GCS_unnamed ellipse",
        DATUM["unknown",
            SPHEROID["Unknown",6370997,0]],
        PRIMEM["Greenwich",0],
        UNIT["Degree",0.017453292519943295]],
    PROJECTION["Lambert_Azimuthal_Equal_Area"],
    PARAMETER["latitude_of_center",45],
    PARAMETER["longitude_of_center",-100],
    PARAMETER["false_easting",0],
    PARAMETER["false_northing",0],
    UNIT["Meter",1]]
NAME: String (40.0)
COUNTRY: String (12.0)
POPULATION: Real (11.0)
CAPITAL: String (1.0)
```

A couple of things to note about the coordinate conversion: First, we didn't do a format conversion—the output is a shapefile, just like the input. Second, we used -t_srs to transform the coordinates to the desired projection. When we look at the results using ogrinfo, we see that indeed the coordinate system was changed to U.S. National Atlas Equal Area (which is a Lambert Azimuthal Equal Area projection).

The default output format for ogr2ogr is ESRI Shapefile so you don't have to specify it with the -f switch when creating a new shapefile.

When would coordinate conversion from the command line be useful? Apart from the reasons we listed at the beginning of this section, suppose you just received a CD containing 300 shapefiles that need to be transformed to a coordinate system compatible with your other data. Using the OGR utilities with a bit of simple shell, Ruby, or Python script would make this a simple and quick task. Loading each layer into a GUI and running a tool to transform the coordinates is fine for one layer, but not for 300.

Raster Conversion

The reasons for doing a raster conversion are pretty much the same as those for vector conversion. Let's add a couple more reasons to the list:

- You want to change the compression type of an image.
- You want to set a no-data value in the image to allow displaying certain areas as transparent.
- You want to rescale (reclass) the pixel values in an image.

In this section, we'll do a few raster conversions to illustrate some of the possibilities available with gdal_translate and gdalwarp.

Extracting Part of a Raster   Let's start by pulling out a piece of the world mosaic raster using a latitude/longitude rectangle. Since I happen to know the coordinates that cover Alaska, we'll use it in our example:

```
$ gdal_translate -a_ullr -180 72 -129 50 -projwin -180 72 -129 50 \
  world_mosaic.tif alaska_mosaic.tif
Input file size is 8192, 4096
Computed -srcwin 0 409 1161 501 from projected window.
0...10...20...30...40...50...60...70...80...90...100 - done.
```

Take a look at that command. The -projwin option specifies the area we want to clip out of the image using the coordinate system. Note we could use the -srcwin option to specify the clip area using pixel coordinates for the upper-left corner and a size in the x and y directions, also in pixels. We also used the -a_ullr option to force the output image to have the bounding coordinates we want; otherwise, it would be offset by a half-pixel in both the x and y directions (see Section 18.2, *Using the Command Line*, on page 332 for more information on the cause of this offset).

To check to see whether this worked, we can open the new file in QGIS or one of the other applications we have discussed. Since we just want to see whether it worked, we could use any a graphic viewer on our system that supports TIFF. In Figure 13.5, on the next page, you can see the result of our effort. You can see from the figure that indeed we cropped out Alaska from the world mosaic.

Figure 13.5: Alaska derived from the world mosaic

Let's run gdalinfo on the new file and examine a few of the details. We'll specify the no-metadata switch to cut down on the amount of output.

```
gdalinfo -nomd alaska_mosaic.tif
Driver: GTiff/GeoTIFF
Files: alaska_mosaic.tif
Size is 1161, 501
Coordinate System is:
GEOGCS["WGS 84",
    DATUM["WGS_1984",
        SPHEROID["WGS 84",6378137,298.257223563,
            AUTHORITY["EPSG","7030"]],
        AUTHORITY["EPSG","6326"]],
    PRIMEM["Greenwich",0],
    UNIT["degree",0.0174532925199433],
    AUTHORITY["EPSG","4326"]]
Origin = (-180.000000000000000,72.000000000000000)
Pixel Size = (0.043927648578811,-0.043912175648703)
Corner Coordinates:
Upper Left  (-180.0000000,  72.0000000) (180d 0' 0.00"W, 72d 0' 0.00"N)
Lower Left  (-180.0000000,  50.0000000) (180d 0' 0.00"W, 50d 0' 0.00"N)
Upper Right (-129.0000000,  72.0000000) (129d 0' 0.00"W, 72d 0' 0.00"N)
Lower Right (-129.0000000,  50.0000000) (129d 0' 0.00"W, 50d 0' 0.00"N)
Center      (-154.5000000,  61.0000000) (154d30' 0.00"W, 61d 0' 0.00"N)
Band 1 Block=1161x1 Type=Byte, ColorInterp=Red
  Mask Flags: PER_DATASET ALPHA
Band 2 Block=1161x1 Type=Byte, ColorInterp=Green
  Mask Flags: PER_DATASET ALPHA
Band 3 Block=1161x1 Type=Byte, ColorInterp=Blue
  Mask Flags: PER_DATASET ALPHA
Band 4 Block=1161x1 Type=Byte, ColorInterp=Alpha
```

The information for the raster confirms that it was assigned the same coordinate system as the source image since we didn't specify otherwise. You can also see the bounding coordinates of the image exactly match those we specified with -projwin. You might also notice we didn't specify an output format. This is because the default output format for gdal_translate is a GeoTIFF. If we wanted to convert the format at the same time, we would have specified it with the -of switch.

GDAL supports more than sixty raster formats, some of which are read-only. Those that are read-write we can be specified as an output format. Let's take a simple example and convert the alaska_mosaic.tif to a PNG:

```
$ gdal_translate -of PNG  -co "WORLDFILE=YES" alaska_mosaic.tif \
  alaska_mosaic.png
Input file size is 1161, 501
0...10...20...30...40...50...60...70...80...90...100 - done.
```

Notice we provided the -co option to create a world file in addition to the PNG. This allows us to use the PNG in a GIS application and have it display in the proper location. When using world files, you have to make sure they stay with the raster file. This is one of the advantages of a GeoTIFF, since it encodes the coordinate system information right in the raster.

Changing the Coordinate System   Let's do another example and change the coordinate system of alaska_mosaic.tif to Alaska Albers Equal Area, a projection commonly used for Alaska data. The EPSG code for the projection is 2964; however, it specifies the units as feet rather than meters. All my other Alaska data is in meters, so for this transformation, we'll go the hard way and specify the projection parameters in proj format:

```
$ gdal_translate -b 1 -b 2 -b 3 alaska_mosaic.tif alaska_mosaic_noalpha.tif
$ gdalwarp -t_srs '+proj=aea +lat_1=55 +lat_2=65 +lat_0=50 +lon_0=-154 +x_0=0 \
  +y_0=0 +ellps=clrk66 +datum=NAD27' alaska_mosaic_noalpha.tif \
  alaska_mosaic_albers.tif
Creating output file that is 1296P x 944L.
Processing input file alaska_mosaic_noalpha.tif.
:0...10...20...30...40...50...60...70...80...90...100 - done.
```

In Figure 13.6, you can see the result of the warping process. Notice we translated the image before the warp using gdal_translate. This was to remove the alpha channel from the original image. The alpha channel is used to determine transparency for each pixel in the image. In the case of our mosaic, it was found by trial and error that the oceans were transparent. When warped, they turned black. To get around this problem, we used gdal_translate with a series of -b switches to specify which bands in the image should be used in the output image. This effectively strips the alpha band. If you look back at the gdalinfo output for alaska_mosaic.tif, you'll see that it reported Bands 1 through 3 as red, green, and blue, respectively, as well as Band 4 as alpha. Once we removed the alpha band, the warp gives us the expected result.

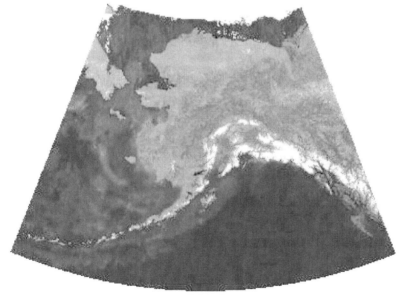

Figure 13.6: Alaska mosaic warped to Alaska Albers projection

Warping an image with gdalwarp is a quick and efficient way to change the coordinate system as opposed to other methods one might use. As you can see, you have to know a little bit about the data you are working with in order to be successful in getting the results you want. If you have more than one image to process

and want to combine them, you can use wildcards with gdalwarp
to mosaic them on the fly, creating a single image in the process.

Raster Transparency   The last raster example we'll look at is setting
a transparency value. This is useful when you want the layers underneath the raster to be visible. To set the transparency, we'll use
the GDAL Virtual Format (VRT). A VRT file is a description of the
raster, stored in XML, that can be modified using a text editor. The
first step is to use gdal_translate to create the VRT file. In this
case, we'll use the survey plat that we digitized from Section 10.1,
Digitizing, on page 137. Running the gdal_translate command on
the plat (ms_724_plat.tif) creates the VRT for us:

```
gdal_translate -of VRT ms_724_plat.tif ms_724_plat.vrt
```

Before we can set the transparency, we have to determine the index
of the white color values in the raster. For this, we use gdalinfo
with the -nomd option since we don't need the metadata, just the
color table information:

```
$ gdalinfo ms_724_plat.tiff
Warning 1: RowsPerStrip not defined ... assuming all one strip.
Driver: GTiff/GeoTIFF
Files: ms_724_plat.tiff
Size is 8144, 6010
Coordinate System is ''
Metadata:
  TIFFTAG_XRESOLUTION=400
  TIFFTAG_YRESOLUTION=400
Image Structure Metadata:
  COMPRESSION=CCITTFAX4
  INTERLEAVE=BAND
  MINISWHITE=YES
Corner Coordinates:
Upper Left  (    0.0,    0.0)
Lower Left  (    0.0, 6010.0)
Upper Right ( 8144.0,    0.0)
Lower Right ( 8144.0, 6010.0)
Center      ( 4072.0, 3005.0)
Band 1 Block=8144x1 Type=Byte, ColorInterp=Palette
  Image Structure Metadata:
    NBITS=1
  Color Table (RGB with 2 entries)
    0: 255,255,255,255
    1: 0,0,0,255
```

From the output, we can see that RGB 255,255,255 (white) is at index 0 in the raster. We now have everything we need to set the transparency using the VRT file. To set the transparency, we add a NoDataValue tag inside the VRTRasterBand tag, using 0 as the index value. Our modified VRT file contains the following:

```
1   <VRTDataset rasterXSize="8144" rasterYSize="6010">
2     <Metadata>
3       <MDI key="TIFFTAG_XRESOLUTION">400</MDI>
4       <MDI key="TIFFTAG_YRESOLUTION">400</MDI>
5     </Metadata>
6     <VRTRasterBand dataType="Byte" band="1">
7       <Metadata />
8       <ColorInterp>Palette</ColorInterp>
9       <NoDataValue>0</NoDataValue>
10      <ColorTable>
11        <Entry c1="255" c2="255" c3="255" c4="255" />
12        <Entry c1="0" c2="0" c3="0" c4="255" />
13      </ColorTable>
14      <SimpleSource>
15        <SourceFilename relativeToVRT="1">ms_724_plat.tiff</SourceFilename>
16        <SourceBand>1</SourceBand>
17        <SourceProperties RasterXSize="8144" RasterYSize="6010"
18         DataType="Byte" BlockXSize="8144" BlockYSize="1" />
19        <SrcRect xOff="0" yOff="0" xSize="8144" ySize="6010" />
20        <DstRect xOff="0" yOff="0" xSize="8144" ySize="6010" />
21      </SimpleSource>
22    </VRTRasterBand>
23  </VRTDataset>
```

The tag we added is found in line 9 of the VRT file. Notice the file also contains a reference to the raster in the SourceFilename. This is important—you still need the original raster in order for the VRT to work properly.

When the VRT file is displayed by software that uses GDAL for raster access, white will be transparent, allowing the underlying layers to show through. In Figure 13.7, on the following page, we have loaded the parcel shapefile we digitized in Section 10.1, *Digitizing*, on page 137 into QGIS, overlaid by our VRT file. You'll notice that the parcels (crosshatch) are visible, proving that the white of our original raster is now transparent. You can use VRT files in a number of ways, including setting transparent values for adjacent images so they can be displayed together without blotting each out.

The key is to use `gdalinfo` to get the color index number and then create and edit the VRT file(s).

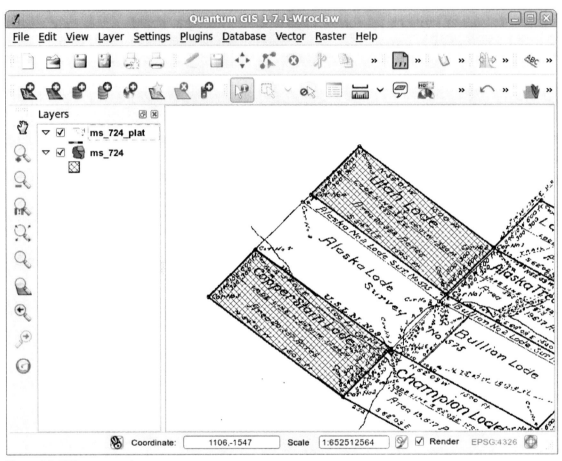

Figure 13.7: VRT raster over the parcel shapefile

We should note that your favorite desktop GIS application likely has the ability to set transparency. For example, QGIS allows you to set custom transparency options on the Transparency tab of the raster properties dialog. In any event, VRT files can be used for a number of interesting things. For more information on the VRT format and what you can do with it, see the GDAL Virtual Format Tutorial.[7]

[7] http://www.gdal.org/gdal_vrttut.html

13.3 *Creating a Spatial Index for Shapefiles*

A spatial index can improve the performance of your mapping application, whether it be on the desktop or the server.[8] The index speeds up drawing, selecting, and identifying features by allowing the software to quickly locate the features of interest. There is more than one way to create a spatial index. In Section 19.1, *Spatial Indexes*, on page 364, you will see how to create one using QGIS. Of course, there is a way to do this from the command line as well. You'll find this useful if you have a whole batch of shapefiles you want to index. Rather than loading each one into QGIS, opening the properties dialog box, and clicking the button to build the index, you can just write a simple script to do the job. To create a spatial index, use the shptree application. You can get shptree in a number of ways. It's included in the FWTools[9] distribution as well as in MapServer.[10] You're probably going to want FWTools anyway, because it contains all the OGR and GDAL utilities we've been using and a lot more goodies. Let's look at the usage:

[8] This assumes that your software supports the use of the spatial index in qix format.

[9] http://fwtools.maptools.org

[10] http://mapserver.org

```
$ shptree
Syntax:
    shptree <shpfile> [<depth>] [<index_format>]
Where:
 <shpfile> is the name of the .shp file to index.
 <depth>    (optional) is the maximum depth of the index
            to create, default is 0 meaning that shptree
            will calculate a reasonable default depth.
 <index_format> (optional) is one of:
          NL: LSB byte order, using new index format
          NM: MSB byte order, using new index format
      The following old format options are deprecated:
          N:  Native byte order
          L:  LSB (intel) byte order
          M:  MSB byte order
      The default index_format on this system is: NL
```

Creating an index is easy, despite all the somewhat confusing options for shptree. In fact, the defaults are usually fine, and you can just use the following:

```
$ shptree earthquakes.shp
creating index of new  LSB format
```

As you can see, there isn't much in the way of feedback. When the command is complete, you'll find a file with a `qix` extension:

```
$ ls -l *.qix
-rw-r--r-- 1 gsherman gsherman 836632 2011-11-09 18:12 earthquakes.qix
```

That's all there is to it. Make sure the spatial index stays with the rest of the shapefile when you copy or move it somewhere else. The spatial index will work with QGIS and MapServer and probably any spatial application that uses OGR for reading shapefiles.

13.4   PostGIS

PostGIS comes with a couple of utilities for moving data in and out of a PostgreSQL database. Although you can accomplish the same results with `ogr2ogr`, the utilities supplied with PostGIS have some additional options that you may find useful. The limitation, of course, is that only shapefiles are supported. Given the flexibility and capability of the OGR utilities, this isn't a problem. We can still get the data from here to there safely and in the form we need.

Importing Shapefiles

To import a shapefile into PostGIS, use the `shp2pgsql` command. Let's look at the options and syntax:

```
RCSID: $Id: shp2pgsql-core.h 5098 2010-01-04 05:47:04Z pramsey
  $ RELEASE: 1.5 USE_GEOS=1 USE_PROJ=1 USE_STATS=1
USAGE: shp2pgsql [<options>] <shapefile> [<schema>.]<table>
OPTIONS:
  -s <srid>  Set the SRID field. Defaults to -1.
  (-d|a|c|p) These are mutually exclusive options:
      -d  Drops the table, then recreates it and populates
          it with current shape file data.
      -a  Appends shape file into current table, must be
          exactly the same table schema.
      -c  Creates a new table and populates it, this is the
          default if you do not specify any options.
      -p  Prepare mode, only creates the table.
  -g <geocolumn> Specify the name of the geometry/geography column
     (mostly useful in append mode).
  -D  Use postgresql dump format (defaults to SQL insert statments.
  -G  Use geography type (requires lon/lat data).
  -k  Keep postgresql identifiers case.
  -i  Use int4 type for all integer dbf fields.
```

```
-I  Create a spatial index on the geocolumn.
-S  Generate simple geometries instead of MULTI geometries.
-W <encoding> Specify the character encoding of Shape's
    attribute column. (default : "WINDOWS-1252")
-N <policy> NULL geometries handling policy (insert*,skip,abort)
-n  Only import DBF file.
-?  Display this help screen.
```

As you can see, shp2pgsql has quite a few options. Let's examine the major ones we need to know in order to get a shapefile into the database. PostGIS uses the concept of a spatial reference ID (SRID), and this option is specified using -s. For the most part, this is equivalent to an EPSG code; however, you can define your own SRIDs in the database. The spatial_ref_sys table contains the spatial reference systems (think projections) that PostGIS knows about as well as the SRID for each. If we look at the record for SRID 4326 in the spatial_ref_sys table, we find that it's the same as EPSG 4326 that we used when doing vector data conversions with OGR. In fact, the authority for that SRID is EPSG.

The -d, -a, -c, and -p switches provide some management options for creating the layer. As you can see, -c is the default option, and you don't need to specify it. The -g switch allows us to specify a name for the column in the table that will hold the feature geometry information. You don't need to specify this unless you don't like the default name or have existing data or standards.

A layer in PostGIS without a spatial index isn't a happy layer. Or at least we won't be happy using it because performance will suffer. You might be thinking we should always specify the -I switch. However, building a spatial index can be a time-consuming affair for large tables. You might want to defer building the index, especially if you are loading a lot of data with a script.

So, what do we really need to specify? Not much when it comes down to it, but here's one word of caution: before you start going wild loading data, think about the SRID you should use and then specify it for every layer.[11] If you don't do it up front, you'll regret it later. Let's load a shapefile as an example:

[11] The shapefiles have to be in the same projection as the SRID you plan to use—loading them into PostGIS won't transform them.

```
shp2pgsql -s 4326 -I cities.shp base_data.cities_pg |psql -d geospatial_desktop
Shapefile type: Point
Postgis type: POINT[2]
SET
SET
BEGIN
NOTICE:  CREATE TABLE will create implicit sequence "cities_pg_gid_seq"
         for serial column "cities_pg.gid"
NOTICE:  CREATE TABLE / PRIMARY KEY will create implicit index
         "cities_pg_pkey" for table "cities_pg"
CREATE TABLE
                   addgeometrycolumn
-----------------------------------------------------------
 base_data.cities_pg.the_geom SRID:4326 TYPE:POINT DIMS:2
(1 row)

INSERT 0 1
INSERT 0 1
...
CREATE INDEX
COMMIT
```

OK, let's analyze what we did here. We knew beforehand that the
EPSG code for `cities.shp` was 4326, so we specified that as the
SRID. The only other option we used was `-I` to build the spatial
index. Other than that, we give the shapefile name and the name of
the table we want to create, in this case `cities_pg`. We prefixed the
table name with the schema name since we wanted our new layer
to reside in the `base_data` schema. If we had ended the command
there and hit `Enter`, we would have seen a slew of SQL statements
scroll by. That's because `shp2pgsql` sends its output to stdout (in
other words, your terminal or command window). We could redi-
rect that to a file using > and then use the file in an application that
was capable of running SQL commands from a file to send it to the
appropriate database. We can take a shortcut, though, and just pipe
the output from `shp2pgsql` directly to `psql`, the PostgreSQL interactive terminal. Passing the name of the database to `psql` using the
`-d` switch is all we need to send the SQL to the database and create
the table.

As the import proceeds, the results are printed to the screen. I
truncated the INSERT statements in the example, since there are

606 of them—one for each city. Notice that before the import began we get some feedback about what's going on, including the creation of the geometry column. If we use `psql` to examine the table after it's loaded, we find that our spatial index was created as part of the import. Partial output from the `\d` command in `psql` shows the GIST index `cities_pg_the_geom_gist` was created on the geometry column.

```
Indexes:
  "cities_pg_pkey" PRIMARY KEY, btree (gid)
  "cities_pg_the_geom_gist" gist (the_geom)
```

Our table is ready to use in our GIS applications; however, you should run the `VACUUM ANALYZE` command to have PostgreSQL collect statistics to improve performance.

Exporting to a Shapefile

To export data from your PostGIS database to a shapefile, use the `pgsql2shp` command. This command has fewer options than its counterpart:

```
RCSID: $Id: pgsql2shp.c 5181 2010-02-01 17:35:55Z pramsey
  $ RELEASE: 1.5 USE_GEOS=1 USE_PROJ=1 USE_STATS=1
USAGE: pgsql2shp [<options>] <database> [<schema>.]<table>
       pgsql2shp [<options>] <database> <query>


OPTIONS:
  -f <filename>  Use this option to specify the name of the file
     to create.
  -h <host>  Allows you to specify connection to a database on a
     machine other than the default.
  -p <port>  Allows you to specify a database port other than the default.
  -P <password>  Connect to the database with the specified password.
  -u <user>  Connect to the database as the specified user.
  -g <geometry_column> Specify the geometry column to be exported.
  -b Use a binary cursor.
  -r Raw mode. Do not assume table has been created by
     the loader. This would not unescape attribute names
     and will not skip the 'gid' attribute.
  -k Keep postgresql identifiers case.
  -? Display this help screen.
```

Basically, we just need to provide the name for the shapefile we want to create and the connection information for the database. You probably realize this means you can use this command from a remote machine. Let's be wishy-washy and export our layer back out of the database:

```
$pgsql2shp -f cities_out.shp geospatial_desktop base_data.cities_pg
Initializing... Done (postgis major version: 1).
Output shape: Point
Dumping: XXXXXXX [606 rows].
```

Things worked as expected—we got all 606 cities out of the database and into a new shapefile. We could use ogrinfo to check it out, but trust me, it's the same as what went into the database. Notice that we didn't specify anything for the database connection. That's because we ran pgsql2shp on the same host as the database server. If we were doing an export from a remote server, we would have to specify host, user, and password. If your database server runs on a nonstandard port, you will have to specify it as well.

If your table has more than one geometry column, you can specify which to use for exporting the features with the -g switch. You might be wondering how you get a table with more than one geometry column. All I'll tell you is it didn't happen with shp2pgsql. Seriously, though, creating a table with more than one geometry column is done programmatically through SQL or custom applications using the programmer's API for PostgreSQL.

That's pretty much the quick tour for getting data out of your PostgreSQL database. There are other ways too, including writing custom scripts using APIs that know how to talk to the database. Although pgsql2shp works well for exporting to shapefiles, you might find that ogr2ogr provides a better solution for exporting to other formats.

14

Getting the Most Out of QGIS and GRASS Integration

We have mentioned the QGIS-GRASS integration in a number of places so far. In this chapter, you will see how QGIS can serve as a front end for viewing and editing GRASS data, as well as for performing analysis and data conversion. We're venturing into some powerful territory here, allowing you to harness the geoprocessing power of GRASS.

QGIS supports GRASS through the use of a plugin. The plugin provides access to GRASS data and functions and is distributed with all official QGIS packages. Actually, it consists of a data provider to bridge between a GRASS map layer and the QGIS map canvas and a plugin to provide the user interface. Well, enough of the boring details; let's get started by loading up the GRASS plugin and seeing how it all works.

To review, when you initially start QGIS, there are no plugins loaded. To load a plugin, you use the Plugin Manager, which can be opened from the Plugins→Manage Plugins... menu. The Plugin Manager provides a list of all the available plugins and whether they are currently loaded (indicated by a checkbox next to their names). To load the GRASS plugin, click the checkbox next to its name, and click the

See Figure 19.3, *Plugins*, on page 375 for a screenshot of the Plugin Manager

OK button. This loads the GRASS plugin, adds a GRASS menu to the Plugins menu, and adds a new toolbar to the GUI as shown in Figure 14.1. For a review of plugins in QGIS, see Section 19.4, *Plugins*, on page 372.

Figure 14.1: The GRASS Plugin Toolbar

If you have followed along with some of the previous GRASS examples, you probably recall how to load the GRASS plugin in QGIS and create a location. If not, refer to Section 18.1, *Location, Location, Location*, on page 323 for a refresher. You might also want to take a look at Section 18.2, *Importing with QGIS*, on page 341 for information on how to import vector layers using the GRASS toolbox in QGIS. We will now expand on those concepts by exploring the toolbox in depth, using it first to load some more data and then to do a bit of data conversion and analysis. From this point on, we assume a working GRASS install and a ready-to-use mapset.

14.1   Loading and Viewing Data

If you work through the GRASS basics in *Appendix C: GRASS Basics*, on page 323, you end up with two vector layers in your mapset: `cities` and `world_borders`. Our goal now is to add the Natural Earth raster to GRASS using the toolbox. We chose the "Cross Blended Hypso with Relief, Water, Drains, and Ocean Bottom" version which is quite large—you may want to start with a smaller version.[1] To get started, we first add our raster to QGIS using the Add Raster Layer tool or menu.

[1] Any of the Natural Earth rasters found on the website will do: http://www.naturalearthdata.com/downloads/10m-raster-data Since the Natural Earth raster had a somewhat cryptic name I renamed it to `natural_earth.tif`.

Once it's loaded up, we need to open our mapset using the `Open mapset` menu item in the `GRASS` menu (remember the `GRASS` menu is located under the `Plugins` main menu). The mapset we want is in our `world_lat_lon` location. In Figure 14.2, on the facing page, you can see the dialog box used to open a mapset in QGIS. Notice that you can open any mapset in any location in any GRASS database using this dialog box. As you change the Gisdbase location, either by typing a path or by browsing to it, the Location and Mapset drop-downs change to reflect what's available.

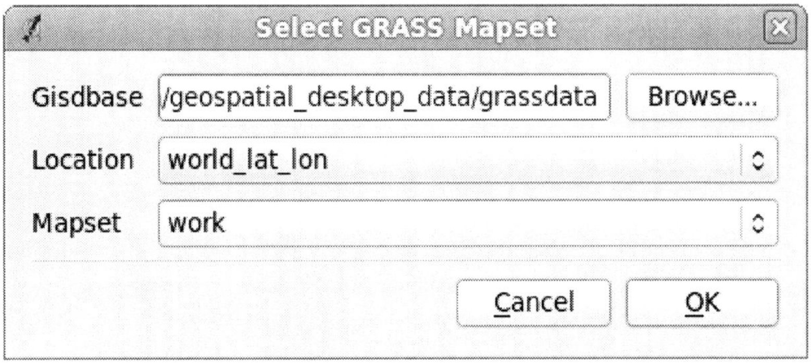

Figure 14.2: Selecting a GRASS mapset in QGIS

Once we open the mapset, the `Open GRASS tools` tool becomes active on the GRASS toolbar, as do the region and vector edit tools. Now that the mapset is open, we can open the GRASS toolbox. Click the `Open GRASS Tools` tool and wait as the toolbox initializes. The tools in the toolbox are added dynamically. In fact, you can customize the tools and add more GRASS functionality. The downside (well sort of) is that the more tools in the box, the longer it takes to open. Once it's up and initialized, you are presented with a collection of tools, as shown in Figure 14.3, on the next page.

Before we import our raster, let's look at the toolbox for a minute. The tools can be viewed as a list or in a tree structure by module. In Figure 14.3, on the following page, you can see the list of modules. Notice the scrollbar in the toolbox—it has a ways to go to get to the bottom. There are a lot of tools in the toolbox and we'll look at some of them a bit later.

To import the raster, we can first filter the module list to find the one we need. If we type "import" in the filter box, we get a list of all modules that deal in some way with importing data—whether it be raster or vector. We want to import a raster so we click the `r.in.gdal.qgis` tool. You must have a raster loaded in QGIS before using this tool, which we do so we are in good shape.[2]

Each tool in the GRASS toolbox has its own page for accepting input and providing feedback as it runs. Typically the tool page will

[2] The same is true for vectors. To import either type using the toolbox, you must load the GDAL/OGR-supported layer in QGIS first. To import a vector, use the `v.in.ogr.qgis` tool.

Figure 14.3: The GRASS tools in QGIS

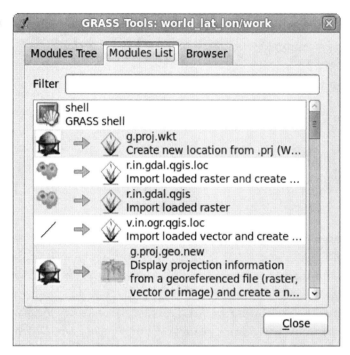

contain Options, Output, and Manual tabs. You enter the required parameters for the tool on the Options tab; then when the tool is run, the output shows up on the Output tab. Clicking the Manual tab displays the manual page for the GRASS tool you are working with.

The Options tab contains a drop-down box populated with the layers eligible for conversion. In this case, since we have only the natural_earth.tif raster loaded, it's the only thing in the list. To convert it, we just need to supply a name for the output map. Rather than be original, we'll use natural_earth for the output name. Now all that's left to do is click the Run button and watch the output fly by. Make sure you wait for the "Successfully finished" message in the output box before proceeding.

Once the raster is imported, we can review the contents of the Output tab to look for any problems or see the results and details of the conversion. Not only that, but it also provides a good way to learn about what's going on inside GRASS.

If you watched closely during the import process you may have noticed that three separate rasters were created: `natural_earth.red`, `natural_earth.green`, and `natural_earth.blue`. These correspond to each band in the original image. If we load one of them into QGIS, we get a grayscale image. We need to composite these to create a representative raster with all the original colors. Although doing this in GRASS is a simple command, we can also do it right from within QGIS. First we need to load each component layer into QGIS. We can easily add these using the Browser in the GRASS toolbox by selecting the raster and clicking the *Add selected map to canvas* button. We also need to make sure the region is set to include our rasters. From the toolbox, click on the Browse tab and expand the raster node. Click on one of the three component rasters—it doesn't matter which—and click the `Set current region to selected map` tool.

Now click on the *Modules List* tab in the GRASS toolbox and enter "composite" in the Filter box. The `r.composite` should float to the top. Click it to open the tool—you'll notice that it as prepopulated all the drop-downs except the name of the output map. We need to be careful to make sure that we choose the proper image for the red, green, and blue inputs. You can experiment with the levels if you like but the default should do fine. Scroll down if need be and enter a name for the composite map—we'll use `natural_earth_composite`. Click the Run button and wait for the process to complete. Once done, you can add the new map to QGIS by clicking on the *View output* button. Figure 14.4, on the next page shows the `r.composite` tool ready to run (two of the level input boxes and the output raster map box are not visible).

You can also set the region from the `r.composite` tool by clicking on one of the *Use region of this map* buttons to the right of the map input name.

Assuming all went well, we now have the `natural_earth` raster loaded into GRASS. If you didn't load it using the *View output* button from the toolbox, you can load it up in QGIS using the `Add GRASS raster layer` tool on the GRASS toolbar.

Figure 14.4: The GRASS Toolbox
Ready to Run r.composite

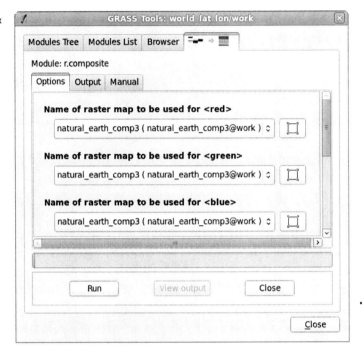

Those are the basics of getting data into GRASS using the import capabilities of the toolbox. If you browse the module list you'll notice that not only can we import rasters, but also other OGR vector layers, including PostGIS. Sometimes it makes sense to convert your data, and other times you can just use them in their native form in QGIS with your GRASS data. If you need to manipulate the data, it's best to bring it into GRASS.

Once you've added a GRASS map to the map canvas, it works pretty much like any other layer in QGIS. For vectors, you can identify features, make selections, view the attribute table, and label features. The one thing you can't do is edit the data—at least not through the editing tools we've seen thus far in QGIS.

14.2 Editing GRASS Data with QGIS

Editing vector GRASS maps in QGIS is also done through the GRASS plugin. The editing tools are designed to work with the underlying

GRASS vector model, which is different from a shapefile or Post-GIS data store. GRASS is topological, meaning it understands the relationship between adjacent features and stores common boundaries only once. Shapefiles and PostGIS data, on the other hand, are nontopological—each feature is stored in its entirety, with no regard for adjacent features.

PostGIS version 2.x (under development) has support for topology. See http://www.postgis.org/documentation/manual-svn/Topology.html.

To be able to edit a GRASS map, you obviously need to have the mapset open and the map loaded into QGIS. You may think I'm stating the obvious here, but in fact you can load a GRASS map *without* opening the mapset. If you do that, the toolbox, region tools, and editing tool will be disabled. Now that everything is in order, we can click the Edit GRASS Vector layer tool, opening the edit tools shown in Figure 14.5.

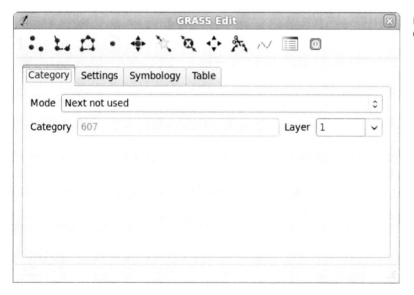

Figure 14.5: GRASS edit tools in QGIS

Which layer are we editing? The answer is the one highlighted in the legend when you bring up the edit tools. Fortunately, if the highlighted layer isn't a GRASS vector map, the Edit GRASS Vector layer tool will be disabled. So, the trick is to make sure the layer you want to edit is highlighted before opening the edit tools.

The edit tools closely mirror those used when digitizing in GRASS, a process you can learn about in Section 18.5, *Digitizing and Editing in GRASS*, on page 351.

First we will look at the basics of the editing tools and then get into the specifics. The task we'll use to illustrate simple editing is adding a new city to the world. Since our mapset is already open, we can load the `cities` layer and use it for our editing task. Each feature in GRASS has a category field named `cat` that serves as the identifier. When we add a new point (city), we have a choice of how that category is created, as shown in the Mode drop-down box. The choices are as follows:

- *Next Not Used*: The next unused category will be assigned automatically.
- *Manual Entry*: You will enter the category manually when you create the feature.
- *No Category*: No category will be assigned to the new feature.

We'll explore some of these options later, but for now, automatically assigning the next available category seems like a good way to go. The first tool on the toolbar is `New point`. Let's use it to create a new city in Alaska named, for lack of a better term, Quantum GIS City. First we zoom in to where we want to create the city, choose the `New point` tool by clicking it, and then click the map to place the feature. When we click, the city is created on the map, and the GRASS Attributes dialog box opens to allow us to enter the information for the new city. In Figure 14.6, on the facing page, you can see the dialog box with the information for our new city (OK, I know that Alaska isn't a country, although some Alaskans wish it were). The category (607) was automatically assigned for us, and we entered the name, country, population, and capital. When we click Update, the information is updated in the GRASS database. We could choose to create another new record or delete this one entirely, in which case our feature would have no attributes.

We added the city, but nothing is saved until we close the editing toolbox. Although I said when you click Update, the attributes are saved, they are actually queued up and ready to be saved. Once we

Figure 14.6: Adding attributes to a GRASS feature

close the editing tools, our changes are saved, and the new city is rendered on the map using the same symbology as the rest of the cities.

Creating Multiple Layers in a Map

One of the features of GRASS is the ability to create more than one layer per map. This means you can have point, line, and polygon layers within a single map. Let's look at a practical example of how this might be useful.

We created Quantum GIS City, and it had a population of one. Now after a major infusion of capital because of it's cottage GIS industry, the population has boomed, and the city has grown by orders of magnitude. But in our GRASS layer, it's just a point. This is where we can take advantage of multiple layers in a map. We can digitize the boundaries of the city, creating either a polygon or line layer. We end up with both the point location and the boundaries and can use one or both at the same time in QGIS, depending on what information we are trying to convey.

To create a polygon for Quantum GIS City, start editing the `cities` GRASS map. This time instead of adding a point, we choose the `New boundary` tool on the toolbar and digitize the boundary of our burgeoning city. Once we close the boundary, the attribute dialog box pops up, allowing us to fill in the attributes. Then we use the `New centroid` tool to add a centroid to the boundary, creating the polygon. Again the attributes dialog box pops up, and we can enter the information again. Now we have three features for our city and attributes for each. Now you are asking yourself, why did we have to enter the attributes each time? The short answer is you don't. Since each of the feature types is for the same city, we should have created the point first, filled in the attributes, and then used its `cat` for the other features. You can do this by changing the mode in the edit toolbox to Manual and entering the category value of the point in the Category field. When we digitize the boundary or add a centroid, it will automatically populate the attribute fields from those associated with the point. This saves time and ensures that the attributes for each feature type are consistent.

Now we have more versatility in our `cities` map, in that we can display points when looking at the worldwide or country view and can display lines or polygons when zoomed in for more detail. By now you should be thinking of all kinds of possibilities this provides, since many "features" may have a point or polygon representation in real life. With shapefiles we would have to create a separate file for each feature type. PostGIS supports geometry collections, but QGIS doesn't know how to visualize them since it makes the assumption of one feature type per layer.

Creating a New Map

What if we want to create a new GRASS vector map and add data to it? We could pop out to the GRASS command line or GUI and do it, or we can use the `Create new GRASS vector` item from the GRASS plugin menu in QGIS. Before you can do this, though, you have to open the mapset where you want the new vector map to live.

The steps involved in creating a new GRASS map are as follows:

1. Open the mapset where the map will reside.
2. Click the `Create a new GRASS Vector` menu item.
3. In the *New Vector Name* dialog box, enter the name for the new map.
4. Click OK.
5. Add features to the map using the edit tools.

When you create a new layer, the edit tools automatically pop up so you can add features. QGIS doesn't add your new map to the canvas yet, so it may be a bit confusing as to what you are doing. Don't worry—the layer will show up once we add a few features and close the *GRASS Edit* dialog box.

When first created, the layer has only one attribute: `cat`. Let's say we are creating a new map named `water_wells`. This map will contain the point locations of water wells, and we will digitize them off a raster map we have showing the locations. Having only the `cat` attribute by itself isn't very useful when it comes to storing information about the wells. We need some additional information, such as owner name, depth, and whether it's an active well. To do this, we use the *Table* tab in the edit dialog box.

In Figure 14.7, on the next page, you can see the table I constructed for the water wells map. To build the table, you just use the *Add Column* button to add a new column and then give it a name and set the type and length as appropriate. You can see in the figure I added the `owner`, `depth`, and `active` fields. For the `active` field, I chose to use a single character that will contain a "Y" or "N" to represent yes or no. Once everything is set up, clicking the *Create/Alter Table* button saves the changes. We can now digitize the wells and add the attributes we need. We'll start by adding a new well within the Quantum GIS City polygon and assigning the owner as the city, depth of 220, and a Y in the `active` column to indicate it is in use. From there, you can go forth in like fashion and digitize all the water wells.

We quickly realize in using our shiny new `water_wells` map that it is lacking something. We should have had a `name` field to store the

Figure 14.7: Adding columns to the new GRASS map table

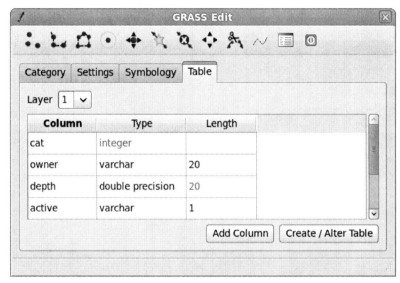

name of the well. For residential wells, this probably isn't important, but it's likely that the managers of Quantum GIS City will want names on their wells to more easily manage them. No problem—we can easily add the new column and fix any existing wells so they have a name.

To add a name column, begin editing the water_wells map, and click the *Table* tab. We now see the table with its current columns and their types. To add a new column, just click the *Add Column* button, and fill in the details.

What about the well(s) we already created? The GRASS edit tools allow you to edit the attributes for any feature. To edit the attributes for an existing feature, click the Edit attributes tool. This brings up the same dialog box you see when entering the attributes for a newly created feature. Now there is a blank name field. We just fill it in with the name of the well, as shown in Figure 14.8, on the facing page, and click the *Update* button. You'll have to repeat this for all the existing wells. Of course, this illustrates the point that you should think about your requirements before creating the data. It will save you time and energy, especially if you realize it too far into your project.

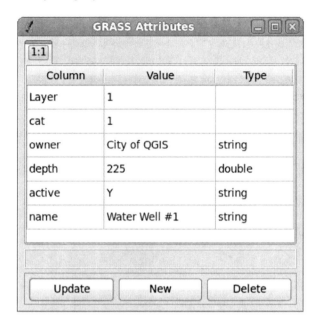

Figure 14.8: Editing the attributes of an existing feature

The Rest of the Story

That pretty much covers the basics of editing GRASS maps with QGIS. Before we move on, let's look at some settings you can customize while editing.

The first is the *Settings* tab. Currently there is only one setting—the snap tolerance. This controls how close you have to be to a feature for the current tool to snap to it. It's set in screen pixels. You can customize it for your use if you find the default value of 10 is too large (or small). Having too large a snap tolerance may result in selecting (and possibly deleting/moving) the wrong feature.

You can also customize the size and colors used for displaying items by clicking the *Symbology* tab. This allows you to customize the line width used when digitizing, as well as the marker size used for points and centroids. Below that is a list of the colors that can be customized for things like background, highlight, point, line, boundary, centroid, and so forth. It's good to have options, but in most cases I find the default colors to be fine for digitizing.

To complete our editing saga, the results of our efforts are shown in Figure 14.9. Each of the GRASS layers (feature types) for the `cities` map is shown in the legend. I've renamed them so you can tell which layer type they represent. We also see the `water_wells` map, containing two wells. You'll notice that the wells are rendered by their status—active or inactive. We can see from the map that the Quantum GIS City fathers have decommissioned old Water Well #1.

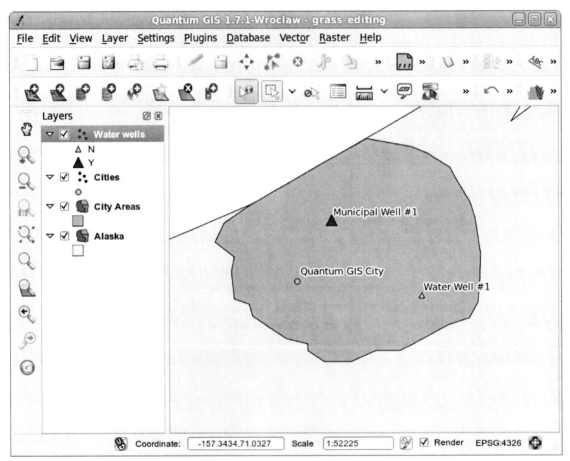

Figure 14.9: Completed city map with water wells

Which method of GRASS editing should you use—QGIS or the

GRASS GUI? It depends mainly on preference, how comfortable you are with the tools, and whether you can perform all the tasks you need to from the chosen interface. In the next section, we will take a look at the analysis and conversion tools available in the GRASS toolbox in QGIS. This may help you decide whether you can do most of your GRASS work in QGIS or whether the GRASS GUI is the way to go.

14.3   Using Analysis and Conversion Tools

Creating and drawing data in GIS is just part of the picture, unless of course all you want is a pretty map. A big part of GIS and what makes it a powerful tool is the ability to do analysis of spatial relationships. In this section, we plan to take a look at some of the tools available in the QGIS GRASS toolbox that allow you to do both conversion and analysis. But before we go there, let's complete our look at the *GRASS Tools* dialog box.

We got our first look at the toolbox back in Figure 14.3, on page 240. We used a couple of the tools to convert some vector and raster data into GRASS but didn't really look at the toolbox that closely. If you refer to the figure, you'll see at start-up there are three tabs: *Modules Tree*, *Modules List*, and *Browser*. The The two module tabs contain all the GRASS tools you can run from the toolbox. These tools are added at runtime from a configuration file, and there is actually a way to customize and add to the tools that are presented in the toolbox, assuming you have attained the appropriate level of GRASS mastery. The tools are categorized by function in the tree and listed in no apparent order in the modules list.

See the QGIS User Manual for information on customizing the toolbox.

We will use some of the tools in the module list shortly. The *Browser* tab contains the GRASS browser and allows you to view all the maps in your current mapset, as well as manage them. Before we dive into the modules, let's learn a bit about the browser.

Using the Browser

To activate the browser, simply click the *Browser* tab. All the mapsets in the current location are displayed on the left in a tree structure. Typically you will see the PERMANENT mapset, along with the one or more user mapsets. Remember, the PERMANENT mapset contains read-only maps that are shared among users and are generally base layers everybody needs.

If a mapset contains maps, you will be able to expand the tree. The maps are further categorized into raster, region, and vector nodes. Expanding one of the nodes will show you a list of all the maps associated with it. If you click a map, the pane on the right displays information about the map. In Figure 14.10, on the next page, you can see the browser with the cities map selected. Notice all the good information displayed in the pane on the right? We can get a good overview of the map, including the number of each feature type (in this case 607 points, 0 lines, 1 boundary, 1 centroid, 1 area, and 1 island). The original layer as imported from the shapefile had 606 points. The other features were added when we digitized the city limits of Quantum GIS City and its point location. We can also see the extents of map—in this case it takes up most of the world.

As we work with GRASS maps, a history is recorded. You can see in our browser example the command used to create the cities map from the original shapefile. This is displayed in the right pane, just below the extent information. Also note on the left under the cities node, there are three layers, one for each feature type. These layers are prefixed with a number, followed by the feature type. Using the browser gives you a quick overview of your maps and layers, as well as some detailed information about the number of features and the history of the map.

Let's take a look at the toolbar in the browser because it has a number of useful functions for working with and managing our maps.

The first tool is the Add selected map to canvas tool. Its purpose

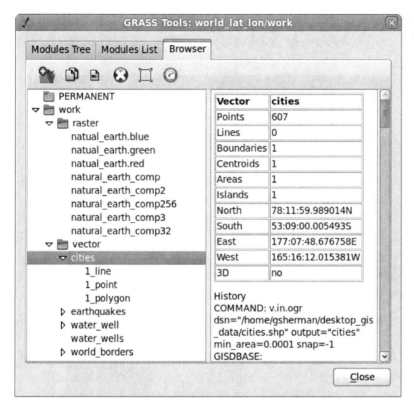

Figure 14.10: GRASS browser in QGIS

seems pretty obvious—you select a map and click the tool, and it gets added to the map canvas in QGIS. What gets added depends on what you have selected in the list of maps and layers. If you have a map selected, for example, `cities`, and click `Add selected map to canvas`, all the layers under `cities` will be added. This means the point, line, and polygon layers get added as separate layers in QGIS, and we will have three entries in the legend. If you had selected just the polygon layer of the `cities` map, only it would be added. This feature gives you a quick way to add every layer for a map or to be more discriminating and add only the particular layer you need.

The next tool is `Copy selected map`, which allows you to copy the currently selected map into the current mapset. There are a number of reasons why you might want to copy a map. A typical case is

when you are about to do some fancy (read: dangerous) conversion
or edit—you might want to make a copy in case things go bad.
When you click Copy selected map, you are prompted for a name
for the new map. Give it a name, and click OK to create the new
map. Note that in order for this to work, you have to highlight a
map in the browser list—the copy won't work on a layer of a map.

The Rename selected map tool allows you to rename a map. This
can come in handy when doing a risky edit operation. Before you
begin make a copy of your map. If things go bad during the edit
operation, you can just delete the original map that is now fouled
up and rename the copy to the original so you can try again (don't
forget to make another clean copy). When naming or renaming
a map, don't think that your spacebar is broken. The prompt for
renaming and copying doesn't allow you to enter spaces since they
are not valid in a GRASS map name.

The red button with the big X in the middle is the Delete selected
map tool. Use this one with caution, because once it's gone, it's
gone. Fortunately, you have to confirm the delete operation, giving
you a chance to change your mind. When you delete a map, it is
removed from the list of maps in the browser. Interestingly enough,
if it happens to be on the QGIS map canvas, it isn't removed, and
you can still identify features and view the attribute table. If you try
to edit it, the operation will fail since the underlying data structures
have been removed.

There are two other buttons on the toolbar. The first allows you to
set the GRASS region to the currently selected map. This information is saved, and the next time you run GRASS, the region will be
restored. Unlike a lot of file system browsers, the GRASS browser
doesn't continuously poll or receive notification when the contents
of your mapset has changed. The remaining button allows you to
refresh the browser contents when you have added new maps or
layers and want to view their information in the browser.

Now that we have a handle on the browser, we'll move on to looking
at some of the basic modules. The browser will come in handy later

when we need to view or manage some of the output maps from our analysis and conversion activities.

Working with Modules in the Toolbox

In previous sections we've seen how to use the import modules in the toolbox to import both vector and raster data. In this section, we'll move beyond that and look at some of the other modules and what you can do with them. First a word about how the modules in the toolbox work: unlike working in GRASS, the modules in the toolbox require a layer loaded in QGIS to use as input to the module. For example, when we imported the `natural_earth.tif` into GRASS, we loaded the TIFF into QGIS first and then used the import tool. Let's start our exploration of the toolbox by creating a buffer or two.

Buffering Vector Features

We can use the toolbox to buffer point, line, or polygon features. Buffers are useful for visualizing "things" within a given distance of other "things." For example, suppose Harrison has spotted eagle nests in an area where a new trail is to be built. He is concerned about the potential for all the eager hikers disturbing the baby eagles and would rather keep them at least 500 meters from the nest. Harrison goes off to help the city planners with a bit of analysis.

With a map of eagle nests, we can create a polygon layer that buffers each nest by a distance of 500 meters. Basically, it's like drawing a circle with a radius of 500 meters around each nest. Once we have the buffer layer, we can use it to site the trail to avoid the nests. On the flip side, you might also analyze a proposed trail by buffering it and seeing whether it overlaps any nests. Of course, this is just a simple example. In practice, buffers are an important part of GIS analysis in many disciplines.

When you buffer a feature, you must specify the distance in map units. In other words, if your map is in latitude and longitude, you would specify the distance in decimal degrees. This usually isn't very practical, so in most cases a projection that uses meters or feet for units of measure is used. Obviously to be successful (and accurate) in your analysis, you have to know a little bit about your data, its projection, and units of measure.

To quickly find a tool in the *Modules List* use the Filter box to type in all or part of the tool name.

To create the buffer, we first put the eagle nests map over the topographic raster map. Then from the GRASS toolbox, we selected the Vector buffer (v.buffer) module from the *Modules List*. When you click the module, it opens a new tab for the buffer operation in the toolbox. Since we had only one GRASS map loaded, the eagles_nest map is chosen as the input vector map. We then just specify the buffer distance in map units, in this case 500 meters, and eagles_nests_500m_buf as the name for the output map.

Figure 14.11: Buffer module ready to buffer eagle nest locations

Figure 14.11 shows the buffer module ready to run. When we click Run, the buffer layer is created, and we can review the output from the buffer processing if we want. To add the newly created buffer to

the map canvas, we just click the *View output* button. It sounds like a lot of steps, but in reality it takes about ten seconds to create a buffer, depending of course on how fast you type.

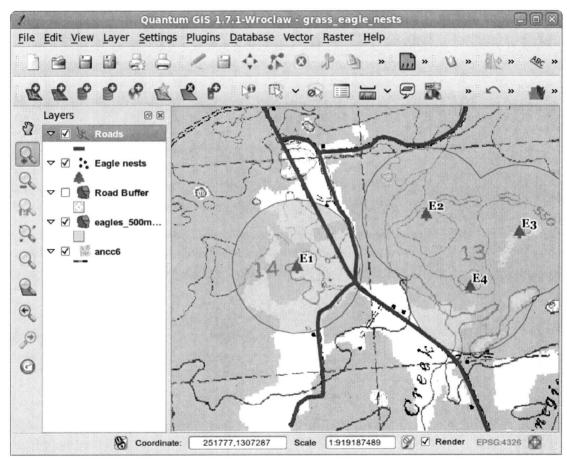

Figure 14.12: Buffered eagle nests created with GRASS and QGIS

In Figure 14.12, you can see the results of the buffer operation. The nests and the buffers are displayed over a topographic map. In this case, we are interested in the distance of each nest from the roads in the area. From looking at the map, we can see that all of the nests except E1 are at least 500 meters from the nearest road. Nest E4 comes close to being within the buffer distance of the road but is still at least 500 meters away. If we were doing an impact analysis, this simple tool provides a quick way to visualize the relationships.

Let's help the eagles shop for a new nest site. In this case, we need to buffer the roads by 500 meters to define the areas unsuitable for nesting. We first start by creating a new empty roads map using our QGIS-GRASS skills. Then we digitize the main roads from the topographic raster map using the GRASS edit tools. Since this is a one-shot analysis, we can get away without entering attributes for our roads. If we were going to use the road map in future work, we would need to put a little more thought into the attributes and enter them as we digitize.

Once we have the roads digitized, we can use the Vector Buffer module to buffer the roads. When we display the roads and the buffer, we can easily see sites in Figure 14.13, on the facing page that aren't suitable for marketing to Mr. and Mrs. Eagle.

You can access the GRASS manual from the GUI Help menu or by typing g.manual -i from the GRASS command-line.

You can also buffer a raster map, in which case the cells of the raster are buffered based on distance zones you set up. If you are interested in this kind of buffer processing, the GRASS manual is your friend.

You can see the utility of a simple geoprocessing task like buffering. With the GRASS plugin in QGIS, this kind of GIS analysis is easy to do.

Vector Overlays

Now we turn our attention to a group of modules that have to do with vector overlays. These modules can be found on the *Modules List* tab of the toolbox using "overlay" as the filter. The modules available to us include the following:

Vector union

This module overlays two vector maps to create a new map containing the union or combination of the features. The attribute tables from the two maps are merged, and a prefix is added to column names so you can tell from which map they originated.

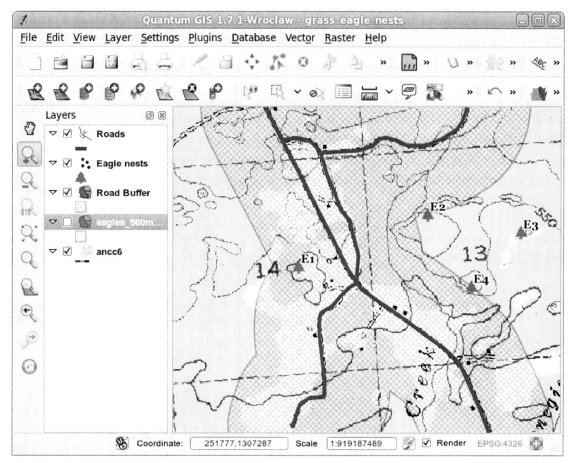

Figure 14.13: Roads buffered using GRASS in QGIS

The boundaries are not dissolved so all original boundaries of the features are still visible.

Vector intersection

This module creates a new map containing the portions of features common to features on the input maps. Where two polygons overlap, only the common portion will be included in the resulting map.

Vector subtraction

The second polygon map is subtracted from the first. The result for each feature is that portion of the feature in the first map not overlapping a polygon feature in the second map.

Vector non-intersection

This module removes the common portion between polygon features. If two polygons overlap, the overlapping part will be removed, resulting in what looks like a hole in the combined polygons.

In Figure 14.14, you can see a graphic illustration of the results of each vector overlay operation. In the center are two polygons, and surrounding them is the result. The graphics in the GRASS toolbox also provide a visual cue for each overlay operation in case you need reminding. You can start thinking about how you would use these modules for something real. We'll take a simple example to illustrate the use of the Vector subtraction command (v.overlay.not). This will be sufficient to illustrate the use of the modules, and then you can go crazy with the others.

Figure 14.14: Result of each type of vector overlay operation

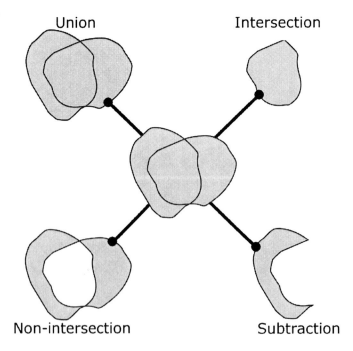

Let's take a semireal but mostly manufactured example using logging.[3] Suppose the logging company is allowed to cut certain stands
of trees, based on species and age. We have a polygon map outlining
the stands eligible for cutting. A stream runs through the logging
area. To protect both the stream and the fish populations, there is a
100-meter setback requirement from any activity. We need to identify the portions of the eligible stands that are "legal" for harvest.

First we prepare our data and make sure we have a good and accurate polygon map of the eligible stands. Next we need to buffer
the stream to 100 meters to create the second polygon map needed
for the analysis. Once we have those, we can proceed with the subtraction operation. In Figure 14.15, on the following page, you can
see the stands map and the digitized streams along with the buffer.
You can guess by looking at it where the likely "no logging" areas
are, but by doing the analysis we will be able to visualize it to aid
in making decisions.

Now to subtract the portions of the stands that are not eligible for
logging. To do this, we use the Vector subtraction module and enter the name of the stands map as the first vector input map on
the module input screen. The stream buffer map is entered as the
second vector map. Then we specify a name for the output map.
Since it will contain polygons of the eligible areas, we'll name it
`eligible_stands`. We click the Run button, and off we go. The
result of this operation is shown in Figure 14.16, on page 263.

You can see from the results that the upstream stand has been
carved up into three fairly small pieces, one of which is between
a fork in the stream. If we pretend to know something about logging, we might say that the upstream stand (the one to the right)
doesn't look like it's too promising in terms of both size and location. The downstream stand has been cut in two but is still fairly
sizeable. This example serves to show how the vector overlay modules can be used for visual analysis of spatial relationships. OK,
enough pretending that I know anything about the timber industry.

[3] Before you take this example seriously, I confess I know little about timber industry practices—it's just an example.

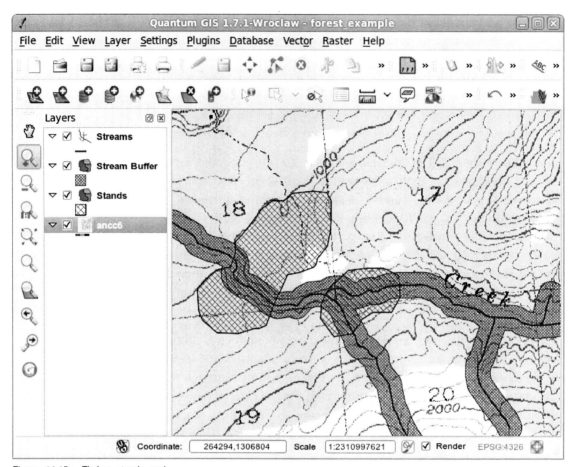

Figure 14.15: Timber stands and stream buffers

Creating a Contour Map

To further illustrate the power of the GRASS modules in the toolbox, we'll create a contour map from the Anchorage DEM we used in the LOS analysis in Section 12.2, *Line-of-Sight Analysis*, on page 175. Creating a contour map is quite simple, but you need to be aware of the limitations. When using a DEM, the map will be only as good as the original data. If the cell size is 50 meters, you can't expect to create contours at 20 meters. The same holds true for any raster source we might want to use.

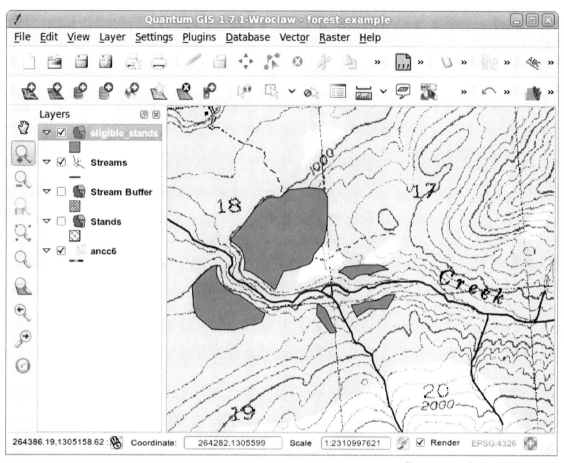

Figure 14.16: Eligible logging areas after vector subtraction

To create a contour map, we first fire up QGIS and open the GRASS mapset that contains the DEM. To make a contour map, we have to add the DEM to the map canvas. Once that's done, we can open the GRASS toolbox and locate the r.contour module.

When you click r.contour, the module *Options* tab is displayed, as shown in Figure 14.17, on the next page. Here we have filled in the parameters for creating the contour map. The first step is to select an eligible GRASS raster from the *Name of Input Raster Map* drop-down. We'll contour the ancc6_dem6 DEM at an interval of 200 feet. Since the DEM is in meters, we need to convert that to feet. Using 3.28 ft/m gives us roughly 61 meters, which we entered in the increment field. The *Minimum contour level* setting specifies how "low" we want to contour. In the case of our DEM, we think everything is above sea level, so we just leave that set at zero. We put a large value for the *Maximum contour level* to make sure we caught everything. The only other thing to specify is the name for the output map.

Figure 14.17: Setting up to contour a DEM

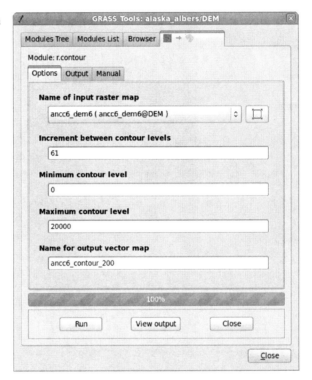

When we click Run, the magic happens, and the vector contour map is created. We can then add it to the map using the *View output* button. The result is shown in Figure 14.18, on the facing page, draped over the original topographic map and the shaded relief. You may find you need to adjust the parameters to get the result you want. If so, you can use the browser to delete the contour map and start over. Make sure you remove the map from the QGIS canvas first!

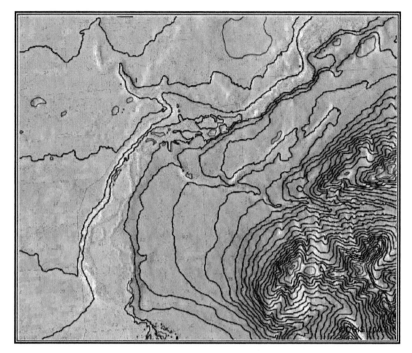

Figure 14.18: Result of contouring the DEM

Map Algebra

If you've been following along with QGIS and actually opened the toolbox, you're probably thinking "Wow, there's a lot of modules in there." You're right, and there is no way we are going to give you an example of each. The goal is to get you started with the basics, and you can develop your skill set to meet your needs. That said, there is one last module we want to look at, just because it's a bit different from the others.

Map or "grid" algebra allows you to perform operations on raster maps in GRASS. This can be useful for a number of things, depending on your data. You might recall that we have already used some map algebra (r.mapcalc) in Chapter 12, *Geoprocessing*, on page 171 when processing rasters.

The QGIS-GRASS toolbox includes a graphic means to design a set of operations to create a new raster from a set of input maps. Essentially you are creating a model that can be run to perform the operation(s). To illustrate, we'll convert our DEM from meters to feet, a simple matter of multiplication.

To convert the DEM, the value of each cell in meters must be multiplied by 3.28 to convert it to feet. Given that our simple little DEM contains thousands of cells, this is no trivial matter. Fortunately, the r.mapcalc module makes this easy to do.

First we'll look at a complete "model" and then explain the process of putting it together. In Figure 14.19, on the facing page, you can see the completed model ready to run.

So, how did we build the model? Basically, it's a select-drag-drop operation for each component. The tools on the toolbar allow you to add a map, constant, function, and connector. We started out by adding the DEM, which must already be loaded into QGIS; otherwise, it won't show up in the list of available maps. We then added the constant 3.28 and a multiply operator. The output "widget" was already on the model when we started. Once all the parts are in place, we just use the Add connection tool to connect them, making sure they are in the proper sequence. The last step is to enter a name for the output map, and then we are ready to run it.

When we run the model, it's actually just building up a GRASS r.mapcalc command for us in the background and executing it. It multiplies each cell in the ancc6_dem by 3.28 and stores the value in the output map. When complete, we have a new raster map that looks just like the original when displayed in QGIS, but the elevation units are in feet rather than meters. If you think you may want to run the model again, you can save it for future use by clicking the Save tool in the toolbar. When it comes time to run the model again, start the r.mapcalc module, and load the model using the Open tool on the toolbar.

This simple example illustrates how to use the r.mapcalc module to build a model and run it. We didn't look at all the functions, but there are twenty-five operators (arithmetic and logical) and thirty-

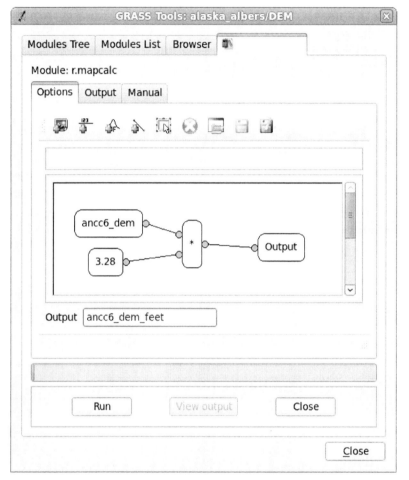

Figure 14.19: Mapcalc model for converting DEM from meters to feet

three functions, including trigonometric, logical, log, and others. So, you can get a lot fancier than we did with the r.mapcalc module.

14.4   Summing It Up

That concludes our tour of the QGIS-GRASS plugin and toolbox. As you may have guessed, we just scratched the surface here. It's up to you to explore further and see what other magic awaits you. As I said at the beginning of this chapter, we're bordering on advanced territory here. Keep in mind as you explore the QGIS-GRASS modules that there may be extra options and "power" available from the GRASS shell.

What Options Are There for Printing Maps?

At some point you are going to want to print from your OS-GIS application. If you are using QGIS, you can use the map composer to create a map complete with legend, scale bar, and annotations. You can export from the map composer to a PNG or SVG for printing or further processing in a graphics application.

GRASS provides the ps.map module to produce high-quality PostScript output suitable for printing. A text file containing mapping instructions is used as input to ps.map. There are a lot of instruction keywords that can be used in creating your map. For help getting started with ps.map, see the GRASS manual page. You can find examples of ps.map scripts on the GRASS wiki at http://grass.osgeo.org/wiki/Ps.map_scripts.

If you are using GMT to create maps, the output is already suitable for printing or inclusion in other documents. If you are using one of the other OSGIS applications, check the manual for information on creating hard-copy output.

There are advanced uses for many of the modules in the toolbox that are beyond the scope of our discussion. If you are already an advanced GIS user, you are probably picturing them right now. If you are a casual or intermediate user, you may find that the QGIS-GRASS integration provides access to a rich set of tools for performing data import and conversion. Be careful, though—once you start down that path, you may end up as an advanced user with more ideas than time.

The QGIS-GRASS tools lower the barrier of entry into the world of geoprocessing with GRASS. As you progress in your GIS journey, you'll likely find yourself using the GRASS shell to get at even more power and options.

15

GIS Scripting

Most GIS users that I know end up doing a bit of programming, regardless of the software they are using. There is always some little task that is easier done with a script or a bit of code. In this chapter, we'll look at some methods for automating tasks in OSGIS software. You don't have to be a programmer to do a bit of script writing, especially when you can get jump-started by downloading examples and snippets.

The script languages available to you depend on the application you are using. Applications and tools with a command-line interface (CLI) can be scripted with most any language available. Others have bindings for specific languages. Some nonexhaustive examples include the following:

- *GRASS*: Shell, Tcl/Tk, Perl, Ruby, Python
- *QGIS*: Python
- *GDAL/OGR*: Shell, Perl, Ruby, Python
- *PostGIS*: Any language that works with PostgreSQL, including Perl, Python, PHP, and Ruby

Some OSGIS applications even provide bindings that allow you to write a custom application using a language such as Python.

In this chapter, we will explore some of the techniques used with these applications.

15.1 GRASS

Since the real core of GRASS is comprised of CLI applications, it's pretty easy to use most any scripting language to perform tasks. From Perl, Python, Ruby, and Tcl/Tk, you can "call" an application and capture the output. This makes GRASS easy to automate.

Shell Game

What do we mean by a shell? It's a command interpreter provided with your operating system. If you use OS X, Linux, or a Unix variant, you likely have bash, csh, and/or ksh available to you. Windows has cmd, which has its own language and probably isn't going to be real helpful in shell scripting. Check out MSYS (mingw.org) and Cygwin (cygwin.org).

Probably the simplest way to automate GRASS tasks is using the scripting capabilities of your shell. On Linux and OS X, this is a pretty natural thing to do, because both come with a fully capable shell. On Windows, you may have to install a Unix-like shell such as MSYS or Cygwin to be able to accomplish the same results.

15.2 QGIS

Since version 0.9.x, QGIS has included support for scripting with Python. QGIS provides the following options for using Python:

- Use the Python console from which you can run scripts using the objects and methods in the QGIS API.
- Write plugins in Python instead of C++.
- Use PyQt[1] to build complete mapping applications using Python and the QGIS libraries.

[1] http://www.riverbankcomputing.com/software/pyqt/intro

Why would you want to do any of this? You'd be surprised at the things you might dream up. QGIS has been designed to make the libraries easily usable in your own plugins and applications. With the Python bindings, this brings a whole new world of possibility—

from simple plugins to complete applications. There is a large assortment of Python plugins for QGIS. To find out what's available, view the current list from the Plugins→Fetch Python Plugins menu.

We're going to take a look at a simple plugin to help us get an idea of what can be done with the PyQGIS bindings. Using Python to get started is pretty easy, so don't be afraid if you aren't a programmer. Let's start by looking at the console.

The Python Console

The console is a bit like using Python from the command line. It lets you interactively enter bits of code and see the result. This is a good way to experiment with the interface and can actually be helpful when you are writing a plugin or application.

To bring up the Python console, go to the Plugins menu, and choose Python Console. The console looks similar to a terminal or command window. Make sure you read the little tip at the top.

The console is not of much use if we don't know what to enter into it. Let's try a simple example and change the title of the main QGIS window. The iface object provides you with access to the QGIS API. Using it, we can reference the main window and set the title:

```
qgis.utils.iface.mainWindow().setWindowTitle('Hello from Desktop GIS!')
```

In Figure 15.1, on the next page, we can see the result of our little example, with the console in front and the new title showing on the QGIS window behind it.

Changing the title isn't all that useful, but it shows you how to access the interface to the QGIS internals. The console is a good exploratory tool for learning about the QGIS API.

To manipulate the map canvas, we can try:

```
qgis.utils.iface.zoomFull()
```

This will zoom the map to its full extent. Now you're probably wondering how to find out what functions are available. The an-

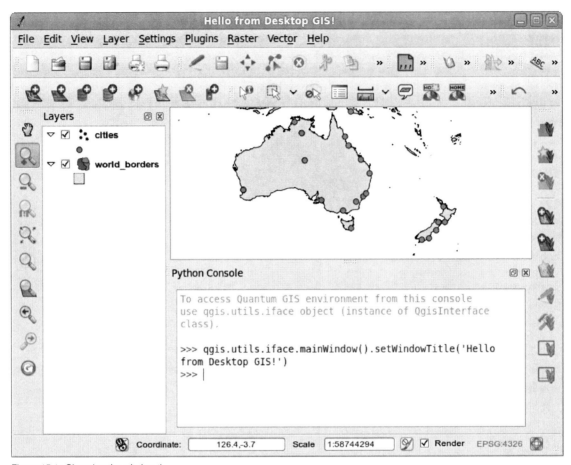

Figure 15.1: Changing the window title with Python

[2] http://www.qgis.org/api

[3] http://www.qgis.org/api/classQgisInterface.html

swer is the QGIS API documentation, available from the website.[2] The API may be a bit intimidating at first, but it's very useful, in fact essential, to our Python exploits. When you use the `iface` object in the Python console, you are actually using an instance of the `QgisInterface` class.[3] If we look at the documentation for `QgisInterface`, we find functions such as the following:

- `zoomFull()`: Zoom to full extent of map layers.
- `zoomToPrevious()`: Zoom to previous view extent.
- `zoomToActiveLayer()`: Zoom to extent of the active layer.

- addVectorLayer(QString vectorLayerPath, QString baseName, QString providerKey): Add a vector layer.
- addRasterLayer(QString rasterLayerPath): Add a raster layer given a raster layer filename.
- addRasterLayer(QString rasterLayerPath, QString baseName=QString()): Add a raster layer given a QgsRasterLayer object.
- addProject(QString theProject): Add a project.
- newProject(bool thePromptToSaveFlag=false): Start a blank project.

We already used the zoomFull method to zoom to the full extent of all the layers on our map. You can see there is a lot of potential here for manipulating the map, including adding layers and projects. We can use these same methods in our plugins and stand-alone applications as well. As you dive into PyQGIS, the documentation will be your friend. You should also keep a copy of the PyQGIS Developer Cookbook[4] handy as well. It contains a lot of information for doing common operations like adding and working with layers, dealing with projections, styling layers.

[4] http://www.qgis.org/pyqgis-cookbook/

Think of the Python console as a workbench for trying methods and using classes in the QGIS API. Once you get that under your belt, you're ready for some real programming. We'll start out by creating a little plugin using Python.

A PyQGIS Plugin

Writing plugins in Python is much simpler than using C++. Let's work up a little plugin that implements something missing from the QGIS interface. For this exercise, you'll need QGIS 1.0 or greater, Python, PyQt, and the Qt developer tools.

Harrison just received the latest *Birding Extraordinaire* magazine, and in it he finds an article that describes locations for the exotic MooseFinch. The locations are in latitude and longitude, which don't mean much to Harrison unless he's in his backyard. He fires up QGIS, adds his layer containing the world boundaries, and begins hunting for the coordinates. Sure, he can use the coordinate display in the status bar to eventually find what he wants, but

MooseFinch: A mythical creature

wouldn't it be nice to be able to just zoom to the coordinates by entering them? Well, that's what our little plugin will do for us (and Harrison).

Before we get started, we need to learn a little bit about how the plugin mechanism works. When QGIS starts up, it scans certain directories looking for both C++ and Python plugins. For a file (shared library, DLL, or Python script) to be recognized as a plugin, it has to have a specific signature. For Python plugins the requirements are pretty simple and, as we'll see in a moment, something we don't have to worry about.

Regardless of your platform, you'll find your Python plugins are installed in the `.qgis` subdirectory of your home directory:

- *Mac OS X and Linux*: `.qgis/python/plugins`
- *Windows*: `.qgis\python\plugins`

For QGIS to find our Python plugin, we have to place it in the appropriate plugin directory for our platform. Each plugin is contained in its own directory. When QGIS starts up, it will scan each subdirectory in our plugin directory (for example, `$HOME/.qgis/python/plugins` on `Linux` and `Mac`) and initialize any plugins it finds. Once that's done, our Python plugin will show up in the QGIS plugin manager where we can activate it just like the other plugins that come with QGIS. You can also specify an additional path for QGIS to search for plugins by using the `QGIS_PLUGINPATH` environment variable. OK, enough of that, let's get started writing our plugin.

Setting Up the Structure

Boilerplate: standardized pieces of text for use as clauses in contracts or as part of a computer program

Back in the old days (around version 0.9) we had to create all the boilerplate for a Python plugin by hand. This was tedious and basically the same for each plugin. Fortunately that's no longer the case—we can generate a plugin template using the Plugin Builder.

The Plugin Builder is itself a Python plugin that takes some input from you and creates all the files needed for a new plugin. It's

then up to you to customize things and add the code that does
the real work. To use the Plugin Builder, first install it from the
Plugins→Fetch Python Plugins menu as seen in Figure 15.2.

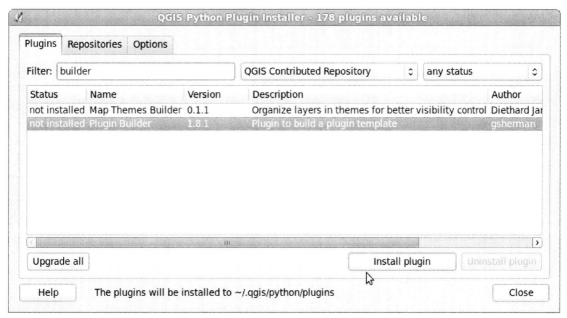

Figure 15.2: Installing the Plugin Builder

When you click on the *Install plugin* button the Plugin Builder will
be installed and you'll find a tool for it on the Plugins toolbar and
menu entries under Plugins→Plugin Builder

Let's generate the structure for our Zoom to Point plugin by clicking on the Plugin Builder tool or menu item. We are presented
with a dialog that contains all the fields needed to create the plugin.
On the left side of the plugin dialog you'll see some hints about
what is expected for each field. Figure 15.3,on the next page shows
all the fields needed to generate the Zoom to Point plugin.

When we click on OK, the Plugin Builder generates a bunch of files:

icon.png
    A default icon used on the toolbar. You will likely want to cus-
    tomize this to better represent your plugin.

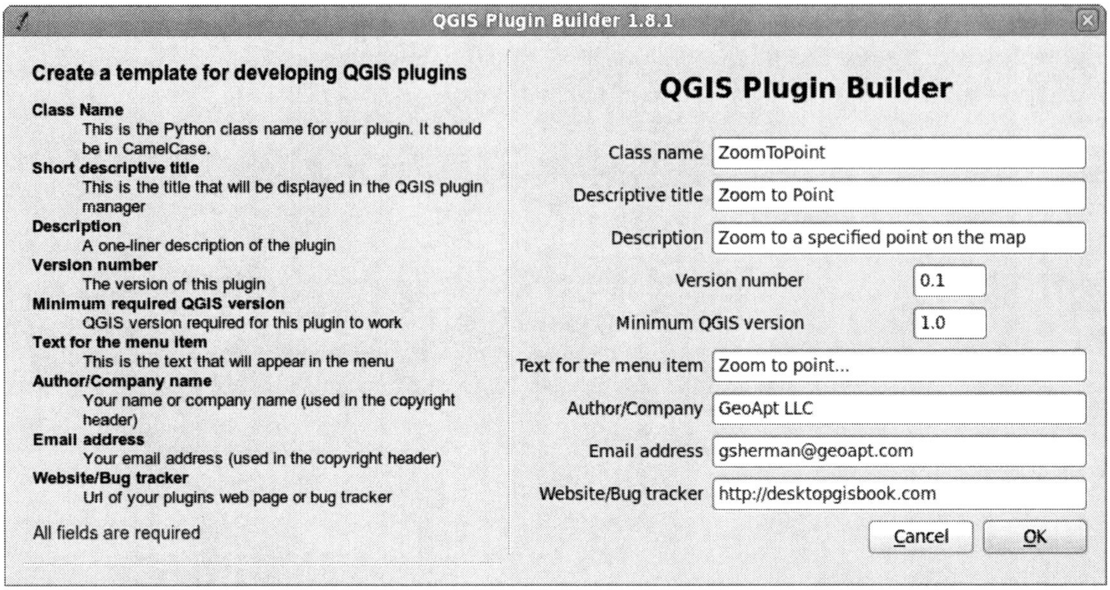

Figure 15.3: Plugin Builder Ready to
Generate the Zoom to Point Plugin

__init__.py

This script initializes the plugin, making it known to QGIS. The
name, description, version, icon, and minimum QGIS version are
each defined as Python methods.

Makefile

This is a GNU makefile that can be used to compile the resource
file `resources.qrc` and the user interface file (`.ui`). This requires
gmake and works on both Linux and Mac OS X and should also
work under Cygwin on Windows.

metadata.txt

The metadata file contains information similar to the methods in
`__init.py__`. Beginning with QGIS 2.0, the metadata file will be
used instead of `__init.py__` to validate and register a Python
plugin.

resources.qrc

This is a Qt resource file that contains the name of the plugin's icon.

ui_zoomtopoint.ui

This is the Qt Designer form that provides a blank dialog with OK and Cancel buttons. It's up to you to customize this to build your plugin's user interface.

zoomtopoint.py

This is the main Python class for your plugin that handles loading and unloading of icons and menus, and implements the run method that is called by QGIS when your plugin is activated. You'll need to customize the run method to make the plugin do its magic.

zoomtopointdialog.py

This Python class contains the code needed to initialize the plugin's dialog.

You'll notice the naming of a number of the files is based on a lower case version of the name you provide for your plugin, in this case `zoomtopoint`.

After the plugin generates the needed files, a results dialog is shown that contains some helpful information, as shown in Figure 15.4, on the following page.

The results dialog contains helpful information including:

- Where the generated plugin was saved
- The location of your QGIS plugin directory
- Instructions on how to install the plugin
- Instructions on how to compile the resource and user interface files
- How to customize the plugin to make it do something useful

At a minimum the only files we have to modify to get the plugin functional are the user interface file (`.ui`) and the implementation file (`zoomtopoint.py`). If you need additional resources (icons or images) you will need to modify `resources.qrc`.

Figure 15.4: Results of Generating the ZoomToPoint Plugin

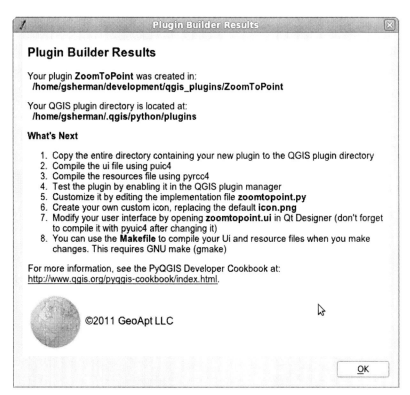

Defining Resources

If you use the Plugin Builder you don't have to modify the resources file, but you do have to compile it. Let's take a look at what's in the resource file:

```
<RCC>
    <qresource prefix="/plugins/zoom_to_point" >
        <file>icon.png</file>
    </qresource>
</RCC>
```

This resource file uses a prefix to prevent naming clashes with other plugins. It's good to make sure your prefix will be unique—usually using the name of your plugin is adequate.[5] The icon is just a PNG image that will be used in the toolbar when we activate our plugin. You can create your own PNG or use an existing one. The only real requirement is that it be 22-by-22 pixels so it will fit nicely on the

[5] Plugin Builder created the prefix for you based on the plugin name.

You don't have to change the icon during development—the default created by Plugin Builder works fine.

toolbar. You can also use other formats (XPM for one), but PNG is convenient, and there are a lot of existing icons in that format.

Once we have the resource file built, we need to use the PyQt resource compiler to compile it:

```
pyrcc4 -o resources.py resources.qrc
```

The `-o` switch is used to define the output file. If you don't include it, the output of `pyrcc4` will be written to the terminal, which is not really what we're after here. Now that we have the resources compiled, we need to build the GUI to collect the information for ZoomToPoint.

Customizing the GUI

Plugin Builder created the GUI for us, but it needs to have controls added to it in order to get our plugin working. To do this, we'll use the same tool that C++ developers use: Qt Designer. This is a visual design tool that allows you to create dialog boxes and main windows by dragging and dropping widgets and defining their properties. `Designer` is installed along with Qt, so it should be already available on your machine.

Our dialog box is pretty simple. In Figure 15.5, on the next page, you can see the dialog box in `Designer`, along with the widget palette and the property editor. It's already complete, but let's take a look at what we had to do to build it. It's going to be a quick tour since we won't go into all the intricacies of `Designer`. If you want to get into the nitty-gritty, see the excellent documentation on `Designer` on the Qt website[6] or in your Qt documentation directory. [6] http://doc.qt.nokia.com

We start by opening our generated dialog box using the `File` menu and selecting `ui_zoomtopoint.ui`. Then we add text labels and text edit controls, as shown in Figure 15.5, on the following page. We also added a spin control for scaling the view. You don't have to set any properties of the text edit controls, but it can be convenient to name them something other than the default. In this case, I named them `xCoord`, `yCoord`, and `spinBoxScale`. This makes it easier to

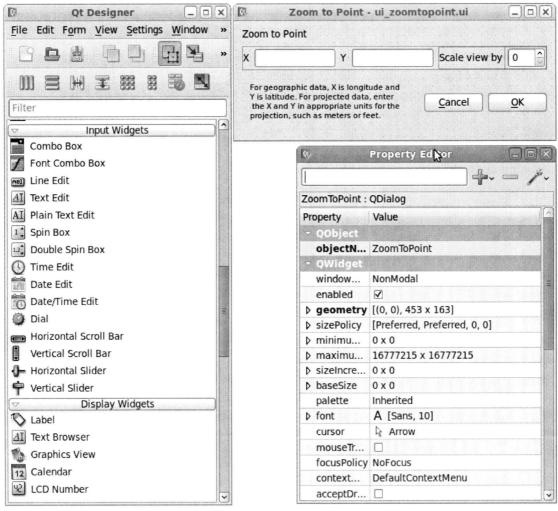

Figure 15.5: Plugin dialog box in Qt Designer

reference them in the code (for those of us with short memories). For our dialog box, we don't need to change the default actions of the OK and Cancel buttons. Once we have all the controls on the form, we're ready to generate some code from it.

To convert our completed dialog box to Python, we use the PyQt `pyuic4` command to compile it:

```
pyuic4 -o ui_zoomtopoint.py ui_zoomtopoint.ui
```

This gives us `ui_zoomtopoint.py` containing the code necessary to create the dialog box when the plugin is launched. It's important to maintain the naming convention as the generated Python code relies on specific names in order to find the components it needs. For example, the Python script that initializes the dialog imports `ui_zoomtopoint.py`:

```
from ui_zoomtopoint import Ui_ZoomToPoint
```

If the compiled dialog is not named properly the plugin will fail to initialize.

Our GUI is now ready for use. All we need to write now is the Python code to interact with the QGIS map canvas.

Getting to Zoom

Up to this point we've been laying the groundwork for our plugin. Before we write the code to zoom the map, let's take a brief look at some of the requirements for our plugin that are found in `zoomtopoint.py`. The Plugin Builder generates this code for us but it is good to get an idea of what it does.

Every Python script that uses the QGIS libraries and PyQt needs to import the QtCore and QtGui libraries, as well as the QGIS core library. This gives us access to the PyQt wrappers for our Qt objects (like our dialog box) and the QGIS core libraries. Here are the first few lines from `zoomtopoint.py`, excluding the header comments that appear at the beginning of the file:

zoomtopoint.py

```
1   # Import the PyQt and QGIS libraries
2   from PyQt4.QtCore import *
3   from PyQt4.QtGui import *
4   from qgis.core import *
5   # Initialize Qt resources from file resources.py
6   import resources
7   # Import the code for the dialog
8   from zoomtopointdialog import ZoomToPointDialog
9
10  class ZoomToPoint:
11    ...
```

You can see that in lines 2 through 4 we import the needed Qt libraries as well as the QGIS core library. Following that we import the resources that contain our icon definition and in line 8 we import the dialog loader class. Line 10 begins our class definition for the plugin. The implementation of our plugin all takes place within the ZoomToPoint class. The methods we are about to discuss are all members of ZoomToPoint and are shown in the listing below:

zoomtopoint.py

```
10  class ZoomToPoint:
11
12      def __init__(self, iface):
13          # Save reference to the QGIS interface
14          self.iface = iface
15
16      def initGui(self):
17          # Create action that will start plugin configuration
18          self.action = QAction(QIcon(":/plugins/zoomtopoint/icon.png"), \
19              "Zoom to point...", self.iface.mainWindow())
20          # connect the action to the run method
21          QObject.connect(self.action, SIGNAL("triggered()"), self.run)
22
23          # Add toolbar button and menu item
24          self.iface.addToolBarIcon(self.action)
25          self.iface.addPluginToMenu("&Zoom to point...", self.action)
26
27      def unload(self):
28          # Remove the plugin menu item and icon
29          self.iface.removePluginMenu("&Zoom to point...",self.action)
30          self.iface.removeToolBarIcon(self.action)
31
32      # run method that performs all the real work
33      def run(self):
```

```
34        # create and show the dialog
35        dlg = ZoomToPointDialog()
36        # show the dialog
37        dlg.show()
38        result = dlg.exec_()
39        # See if OK was pressed
40        if result == 1:
41            # do something useful (delete the line containing pass and
42            # substitute with your code
43            pass
```

When the class is first instantiated, we store the reference to the
iface object using the __init__ method. This method gets called
whenever we create a ZoomToPoint object. We store iface as a class
member because we are going to use it later when we need access
to the map canvas.

As far as QGIS is concerned, plugins must implement only two
methods: initGui and unload. These two methods are used to
initialize the user interface when the plugin is first loaded and clean
up the interface when it's unloaded. Let's take a look at what we
need to initialize our plugin GUI.

First we need to create what's called an *action*. This is a Qt object of
type QAction. It's used to define an action that will later be used on
a menu or a toolbar. On line 18, we create our action by supplying
three arguments:

- The icon for the toolbar. This is a combination of the prefix
  (/plugins/zoom_to_point) and the icon file name (icon.png) as
  specified in our resources file.
- Some text that's used in the menu and tooltip, in this case "Zoom
  To Point plugin."
- A reference to the parent for the plugin, in this case the main
  window of QGIS.

On line 21, we do one last thing with the action to connect it to
the run method. This basically connects things so that when the
OK button on the dialog box is clicked the run method in our
ZoomToPoint class is called.

Next we need to actually put our nicely configured action on the menu and toolbar in the GUI. The `QgisInterface` class that we played with in the Python console contains the methods we need. On line 24, we use `addToolBarIcon` to add the icon for our tool to the plugin toolbar in QGIS. To add it to the menu, we use `addPluginToMenu` method, as shown on line 25. Now our GUI is set up and ready to use.

The `unload` method is pretty simple. It uses the `removePluginMenu` and `removeToolBarIcon` methods to remove the menu item and the icon from the toolbar. Remember this method is called only when you unload the plugin from QGIS using the Plugin Manager.

Finally, we are ready to add the bit of code that does the real work. Like most GUI applications, the bulk of the code has to do with the user interface while a few bytes do the actual work. In the previous listing the run method is shown as it was generated by Plugin Builder. In the listing below we have modified the run method to zoom to a point.

zoomtopoint.py

```
32    # run method that performs all the real work
33    def run(self):
34      # create and show the ZoomToPoint dialog
35      dlg = ZoomToPointDialog()
36      dlg.show()
37      result = dlg.exec_()
38      # See if OK was pressed
39      if result == 1:
40        # Get the coordinates and scale factor from the dialog
41        x = dlg.ui.xCoord.text()
42        y = dlg.ui.yCoord.text()
43        scale = dlg.ui.spinBoxScale.value()
44        # Get the map canvas
45        mc=self.iface.mapCanvas()
46        # Create a rectangle to cover the new extent
47        extent = mc.fullExtent()
48        xmin = float(x) - extent.width()  / 200 * scale
49        xmax = float(x) + extent.width()  / 200 * scale
50        ymin = float(y) - extent.height() / 200 * scale
51        ymax = float(y) + extent.height() / 200 * scale
52        rect = QgsRectangle( xmin, ymin, xmax, ymax )
53        # Set the extent to our new rectangle
54        mc.setExtent(rect)
```

```
55      # Refresh the map
56      mc.refresh()
```

The first step is to create the dialog box on line 35, show it, and then call the exec_ method. This causes the dialog box to show itself and then wait for some user interaction. The dialog box remains up until either the OK or Cancel button is clicked. Once a button is clicked, we test to see whether it was the OK button on line 39. If so, we are then ready to zoom the map.

First we have to retrieve the x and y coordinates and the scale that you entered on the dialog box (lines 41 through 43). We store these in local variables just to make the next step a bit more readable in the code. Once we have the user inputs, we fetch the extent rectangle from the map canvas using the fullExtent method (lines 47). extent. In lines 48 through 52 we calculate the new bounds using the scale value and a constant and use them to create a new rectangle. Once we have the rectangle, we are ready to zoom the map by calling the map canvas setExtent method (line 54). To actually get the map to zoom, we call the map canvas refresh method in line 56 and the map zooms to the rectangle we specified. Once complete, our plugin stands by ready for the next request.

Let's summarize the process of creating a plugin. First we generate a template using Plugin Builder. Next we optionally set up our resource file with a custom icon and design the dialog box using Qt Designer. Finally, we implement the run method where the real work of showing the dialog box, collecting the input, and zooming the map takes place. While we stretched out the explanation, there really isn't all that much hand written code involved in making the plugin. In fact, for the ZoomToPoint plugin there are less than 80 lines of actual code, of which we wrote very little by hand.

There are a number of enhancements you could add to the plugin, including the ability to "remember" the x, y, and scale values that you used the previous go. If you got really fancy, you could also figure out how to set a marker at the point after you zoom. Come to think of it, once you add those features, send them to me, and

I'll include them in the next release of the plugin. Just to prove it
works, you can see the plugin and the values we just entered in
Figure 15.6. Behind it you'll see the map zoomed to the coordinates
we specified. Notice the magnifying glass icon on the right of the
Plugins toolbar (just above the layer list). That's the icon I specified
for the final version of the plugin, and it indeed shows up on the
toolbar. If we were to look in the Plugins menu, we would find an
entry for Zoom to Point as well.

Figure 15.6: ZoomToPoint plugin in
use

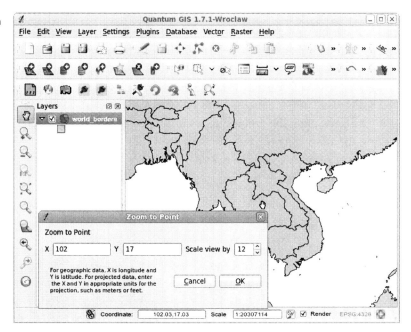

Writing a QGIS plugin in Python is pretty easy. Some plugins won't
require a GUI at all. For example, you might write a plugin that
returns the map coordinates for the point you click the map. Such
a plugin wouldn't require any user input and could use a standard
Qt MessageBox to display the result. You can also write plugins for
QGIS in C++, but that's another story and one I'll let you write.[7]

[7] Actually, you can find information on
writing QGIS plugins in C++ on the
QGIS wiki at http://wiki.qgis.
org.

A PyQGIS Application

A stand-alone application is a step beyond a QGIS plugin. In some ways, they are very similar. We need to create a GUI and use the same imports. On the other hand, we don't have to write all that code to interface with the QGIS plugin mechanism. A stand-alone application does require a lot more GUI coding. Rather than build an application here, I'll point you at a simple tutorial for more information.[8]

[8] http://geospatialdesktop. com/Creating_a_Standalone_ GIS_Application_1

15.3   GDAL and OGR

We already took a long look at the GDAL and OGR utilities in Chapter 13, *Using Command-Line Tools*, on page 199. Here we will take what we learned in that chapter and look at ways to automate our data conversion and loading tasks. Many, if not most, tasks can be handled with a shell script (bash for example) and don't require Ruby or Python. Of course, if you are more comfortable with one of those over shell scripting, it makes sense to use what you know. If you want to tap into the power of the GDAL/OGR bindings, you'll need to use Python, Ruby, Java, or one of the other supported languages.

For Windows users it will likely be easier to use Ruby or Python, unless you have access to a Unix-like shell through Cygwin or MSYS.

In this section, we'll take a look at a couple of examples, one using a shell script and the other using Python with the GDAL/OGR bindings.

Converting Data with a Shell Script

In the simplest case, we have a directory full of files, and we want to perform the same operation on each one. The basic application flow is as follows:

1. Get a list of the files.
2. Loop over the list.
3. Perform some operation on each file.

4. Repeat until all files are processed.

It can't get too much simpler than that. Let's take an example using bash and convert a batch of shapefiles from an Albers projection to geographic coordinates in WGS 84. First let's figure out what command and options we need to get the job done. We'll use ogr2ogr to convert the files. Fortunately, all the shapefiles have an associated projection file, so we don't have to worry about specifying that during the conversion. Here is the bash script to do the conversion:

convert_shapefiles.sh

```
1  #!/bin/bash
2  # Convert all shapefiles in the current directory to
3  # WGS84 projection. The converted shapefiles are placed
4  # in the geo subdirectory.
5  for shp in *.shp
6  do
7    echo "Processing $shp"
8    ogr2ogr -f "ESRI Shapefile" -t_srs EPSG:4326 geo/$shp $shp
9  done
```

Notice that on line 7, we are going to print a little message for each shapefile that is processed. On line 8, we have the ogr2ogr command that does the actual work. As it's processed, each converted shapefile is placed in a geo subdirectory. Other than that, the output from the script isn't that exciting:

```
$ . ./convert_shapefiles.sh
Processing adminbnd.shp
Processing admin_nps.shp
Processing admin_nra.shp
Processing admin_nwr.shp
Processing admin_state.shp
Processing admin_usfs.shp
Processing admin_wild.shp
Processing admin_wild_s.shp
Processing gnisalb.shp
Processing govt_emp.shp
Processing language.shp
Processing owner_fed.shp
```

Let's test one of the new shapefiles to make sure it did what we wanted:

```
$ ogrinfo -al -so adminbnd.shp
```

```
INFO: Open of `adminbnd.shp'
      using driver `ESRI Shapefile' successful.

Layer name: adminbnd
Geometry: Polygon
Feature Count: 11977
Extent: (-168.072393, 53.921043) - (-129.973606, 71.389543)
Layer SRS WKT:
GEOGCS["GCS_WGS_1984",
    DATUM["WGS_1984",
        SPHEROID["WGS_1984",6378137,298.257223563]],
    PRIMEM["Greenwich",0],
    UNIT["Degree",0.017453292519943295]]
AREA: Real (19.3)
PERIMETER: Real (19.3)
BNDS_: Integer (9.0)
BNDS_ID: Integer (9.0)
PARCEL_ID: Integer (9.0)
NAME: String (25.0)
LONGNAME: String (50.0)
ADMIN: Integer (9.0)
AGENCY_ITE: String (50.0)
PARCEL_TYP: String (50.0)
SCALE: Integer (9.0)
EFF_DATE: Date (10.0)
AMEND_DATE: Date (10.0)
DESCRIPTIO: String (50.0)
```

The spatial reference system (SRS) for the converted shapefile is indeed what we asked for—WGS 84.

Creating a Shapefile from Delimited Text

While the GDAL/OGR utilities provide you with a lot of capability, sometimes you may need to dig a little deeper. In this example, we'll use Python with OGR to create a shapefile from the volcanoes dataset. In Section 10.2, *Importing Data*, on page 140, we used the delimited text plugin to import the volcano data into QGIS and display it and then save it to a shapefile. That works well, but suppose you have a lot of data to process. In that case, writing a script to do the work is not only quicker but more flexible.

While ogr2ogr can import CSV files to create a shapefile, this section serves as an example of using the GDAL/OGR Python bindings.

Before we get started, we need to make sure that the Python bindings for GDAL/OGR are present. This is easy to test using the Python Interpreter:

```
$ python
Python 2.7.1+ (r271:86832, Apr 11 2011, 18:13:53)
[GCC 4.5.2] on linux2
Type "help", "copyright", "credits" or "license" for more information.
>>> import osgeo.ogr as ogr
>>>
```

If you get the prompt back with no errors, you are good to go.
If not, it means that your GDAL/OGR install doesn't include the
Python bindings. If you built from source, you'll have to go back
and recompile with the --with-python option. If you don't have
them, a quick way to get the needed bindings for Linux or Windows
is to use FWTools.[9]

[9] http://fwtools.maptools.org

Our script will take the following steps to get from delimited text
to the shapefile:

1. Import the needed modules.
2. Open the delimited text file.
3. Create the shapefile.
4. Add the fields to the shapefile.
5. Read the text file and populate the attributes and geometry
   for each row.
6. Close the shapefile.

Let's take a look at the code in chunks and go through it bit by bit:

import_volcanoes.py

```
1  # import the csv module
2  import csv
3  # import the OGR modules
4  import osgeo.ogr as ogr
5  import osgeo.osr as osr
6  # use a dictionary reader so we can access by field name
7  reader = csv.DictReader(open("volcano_data.txt","rb"),
8      delimiter='\t',
9      quoting=csv.QUOTE_NONE)
```

Beginning with line 2, we import the modules needed for the script.
The csv module is part of Python as of version 2.3. It provides a
simple way to read a delimited text file and is well suited to our

needs. The other imports we need are `ogr`, which provides access to the OGR functions needed to create and write features to a shapefile and `osr`, which provides the spatial reference functions.

Next we set up the `csv` reader in line 7. We are using the `DictReader` class to read the file and map the information into a `dict`. This allows us to reference the data using the field names in the header row of the input file. We'll use this capability to pick and choose which fields we want in our shapefile. When creating the `DictReader`, we need to specify the delimiter, in this case `\t` for the tab character. You might remember we used the same with the QGIS Delimited Text plugin. The last argument in setting up the reader is `cvs.QUOTE_NONE`, which specifies that our file has no quotes around the data.

Now we're ready to use the OGR bindings to create the shapefile:

import_volcanoes.py

```
11   # set up the shapefile driver
12   driver = ogr.GetDriverByName("ESRI Shapefile")
13
14   # create the data source
15   data_source = driver.CreateDataSource("volcanoes.shp")
16
17   # create the spatial reference
18   srs = osr.SpatialReference()
19   srs.ImportFromEPSG(4326)
20
21   # create the layer
22   layer = data_source.CreateLayer("volcanoes", srs, ogr.wkbPoint)
23
24   # Add the fields we're interested in
25   field_name = ogr.FieldDefn("Name", ogr.OFTString)
26   field_name.SetWidth(24)
27   layer.CreateField(field_name)
28   field_region = ogr.FieldDefn("Region", ogr.OFTString)
29   field_region.SetWidth(24)
30   layer.CreateField(field_region)
31   layer.CreateField(ogr.FieldDefn("Latitude", ogr.OFTReal))
32   layer.CreateField(ogr.FieldDefn("Longitude", ogr.OFTReal))
33   layer.CreateField(ogr.FieldDefn("Elevation", ogr.OFTInteger))
```

The first step is to create the driver in line 12. Once we have the driver, we can use it to create the data source. In OGR, a shapefile data source can be a directory of shapefiles, or it can be just a single file. In our case, we are creating just a single shapefile as the data source in line 15. In line 18, we create a `SpatialReference` object and then use the `ImportFromEPSG` method to set the spatial reference to WGS84 using EPSG code 4326.

From the data source, we can create the layer as shown in line 22. The first argument is just the name of the shapefile without the extension. The second argument to `CreateLayer` is the spatial reference. The final argument is the feature type—in our case a `wkbPoint`.

With the layer created, we can add the field definitions in lines 25 through 33. For the text fields, `Name` and `Region`, we create the field object, set an arbitrary width of 24, and then use the `CreateField` to add it to our layer. For the numeric fields, we can create the field all in one step as in line 31.

The layer is now ready for some data:

import_volcanoes.py

```
34   # Process the text file and add the attributes and features to the shapefile
35   for row in reader:
36     # create the feature
37     feature = ogr.Feature(layer.GetLayerDefn())
38     # Set the attributes using the values from the delimited text file
39     feature.SetField("Name", row['Name'])
40     feature.SetField("Region", row['Region'])
41     feature.SetField("Latitude", row['Latitude'])
42     feature.SetField("Longitude", row['Longitude'])
43     feature.SetField("Elevation", row['Elev'])
44
45     # create the WKT for the feature using Python string formatting
46     wkt = "POINT(%f %f)" % (float(row['Longitude']) , float(row['Latitude']))
47
48     # Create the point from the Well Known Txt
49     point = ogr.CreateGeometryFromWkt(wkt)
50
51     # Set the feature geometry using the point
52     feature.SetGeometry(point)
53     # Create the feature in the layer (shapefile)
54     layer.CreateFeature(feature)
55     # Destroy the feature to free resources
56     feature.Destroy()
```

```
57
58    # Destroy the data source to free resources
59    data_source.Destroy()
```

In line 35, we began reading the text file to create the features, one
for each line in the file. For each line we must create a feature object
(line 37) and then set the values for each of the fields in lines 39
through 43. Since we chose to use the DictReader, we can access
the values for each field by name to set the values. This takes care
of the attributes—all that's left is to create the geometry from the
latitude and longitude values.

To create the geometry, we create a WKT representation of the point
using the values of the latitude and longitude fields. In line 46, the
Python string-formatting feature is used to easily create the WKT
for the point in the form of POINT(x y). Using the WKT, the point
feature is created in line 49. The last step to get the entire feature
ready is to set the geometry, as shown in line 52. The feature (attributes and geometry) is now complete and can be added to the
shapefile using CreateFeature, as shown in line 54. The last step
before moving on to the next line in the text file is to destroy the
local feature object to free up resources (line 56).

Once all the lines in the text file are processed, the data source is
"destroyed" to close everything up cleanly. When it's run, the script
produces the following files:

```
volcanoes.dbf
volcanoes.prj
volcanoes.shp
volcanoes.shx
```

If you look at the prj file, you'll see that it contains the projection
information for WGS84. Just to prove it works, the volcano shapefile
displayed over the world_borders shapefile is shown in Figure 15.7,
on the following page.

You can probably envision even more clever and perhaps complicated scripts using both the GDAL/OGR utilities and bindings.
There are GDAL/OGR bindings for Perl, Python, Ruby, Java, and

Figure 15.7: Volcanoes shapefile created with Python script

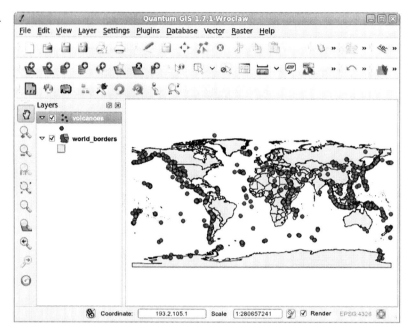

C#/.NET, plus a couple of other languages. This gives you the
option of writing scripts in these languages directly against the
GDAL/OGR API to work with both vector and raster formats.

15.4   Writing Your Own GIS Applications

Most GIS users have ventured into the realm of programming—
whether it be writing scripts as we just looked at building full-blown
applications. Scripting is quite useful for automating GIS tasks but
sometimes you find yourself in a position where you need a customized application.

The full version of your favorite OSGIS application is perhaps overkill
or doesn't provide the features you need. Often trying to twist the
application into the form you need results in a system that is not
user-friendly and is difficult to use. Many disciplines can benefit
from a lightweight custom application that serves a specific need.
These are the reasons for writing such an application; let's look at
some of the specifics.

Options for Writing Your Application

If we are going to write an application, there are some things we *don't* want to build from scratch:

• Low-level drawing routines for displaying raster and vector data
• Read/write access to our data stores
• Renderers such as unique value and graduated symbol
• Legend generation

We want the API to handle all the hard stuff for us so we can concentrate on the custom functionality. That said, let's see what APIs are available for the task.

Tools for Building a Custom Application

You can build a custom application in many ways. You can program at a low-level against, for example, the GDAL/OGR or GRASS libraries. One of the most time-consuming aspects of creating an application is the user interface. If you've never done it before, you'll find that it takes nearly as much code for the GUI as it does for the logic. If you are going to write a desktop application, the first thing you should decide is which GUI toolkit you will use.

Let's take a look at the GIS toolkits (APIs) available for us to work with. If you're like me, cross-platform support is an important consideration, although depending on the scope and target for the application, building on a single OS may be perfectly acceptable. In the list of toolkits that follow, we'll point out the level of cross-platform support for each. This is not a comprehensive list—there are likely other toolkits out there in the wild.

Mapnik
Mapnik[10] is a toolkit for developing mapping applications using C++ or Python—it runs on Linux, Mac OS X, Windows, and FreeBSD. Mapnik renders its output as an image. You can likely integrate this with whatever GUI toolkit you desire, based on your platform. Supported data formats include shapefiles, Post-

[10] http://mapnik.org

GIS, TIFF, and GDAL/OGR.

MapWinGIS

MapWinGIS[11] is an ActiveX control that you can use with any programming language that supports ActiveX on Windows. This includes Visual Basic, Delphi, and C# and the GUI elements that go along with them.

PyWPS

PyWPS[12] is the "Python Web Processing Service," an implementation of the Web Processing Standard from the Open Geospatial Consortium. PyWPS allows you to write applications using Python and GRASS that work over the Web. Although it isn't a way to write desktop applications, it does use GRASS on the back end to provide powerful geoprocessing capabilities via the Web and is an option depending on your needs.

QGIS libraries

The QGIS[13] libraries support stand-alone application development using both C++ and Python. Operating system support includes Linux, *BSD, Mac OS X, and Windows. If you write an application using the QGIS libraries, you'll also be using the Qt GUI toolkit.

uDig framework

The uDig[14] framework supports cross-platform customization using Java and the Eclipse Rich Client Platform (RCP). The framework can be extended through the use of plugins and GUI customization. Since uDig is based on Eclipse, you'll use the SWT toolkit when developing your own customizations.

When choosing a mapping toolkit make sure it can support the data stores you want to use. In addition, you have choices to make: programming language, GIS toolkit, and GUI toolkit.

In many cases you'll find that the mapping toolkits (for example QGIS or uDig) are tightly coupled with both the programming language and the GUI toolkit. In other words, the GIS toolkit you

---

[11] http://mapwingis.codeplex.com/

[12] http://pywps.wald.intevation.org

[13] http://qgis.org

[14] http://udig.refractions.net

choose is going to dictate the other two components. Regardless of
the approach you take, the key thing is to leverage all the hard work
others have done before you, so you can concentrate on implement-
ing your logic, rather than redesigning the wheel from scratch.

16

Appendix A: Survey of Desktop GIS Software

There are a lot of applications in the OSGIS desktop world. In this chapter, we'll explore some of the major choices available. In our survey, we'll classify applications based on both capability and the underlying language. The programming language behind an application is important because it affects how the application is distributed, is installed, and how easy it is for us to customize.

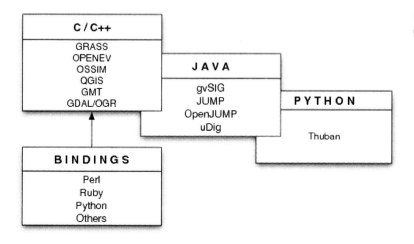

Figure 16.1: Applications grouped by underlying programming language

Sometimes we open source enthusiasts are a bit odd. Many of us choose our software based on the language in which it is written. When you begin to look into OSGIS applications, you will find them divided into what have been termed *tribes* based on the programming language. While we mention the programming language, our focus will center around the features of each application. For reference, in Figure 16.1, on the previous page, you can see the language behind each of the applications in our survey, including the potential for using a scripting language with each. If you are a programmer, you'll be interested in the underlying language since it will give you an idea of how easily you might customize or extend the application. After all, that's a big part of what open source is all about.

Bindings consist of interface code (stored in a shared library or DLL) that allow you to access the features of an application or library from a scripting language.

For most people, the words *Desktop GIS* generally conjure up visions of a GUI interface. Although that's largely true, it's clear there are command-line applications that deserve a place in our toolkit. In the survey, we'll divide the applications into two primary groups—those with a GUI and those that are command line only.

16.1 GUI Applications

Let's start with the GUI applications. A lot of OSGIS applications exist. If you don't believe me, just take a look at the FreeGIS website.[1] We're going to hit some of the major ones. Our survey includes the following apps in alphabetical order:

[1] http://freegis.org

- GRASS
- gvSIG
- Jump/OpenJump
- OSSIM
- OpenEV
- Quantum GIS
- Thuban
- uDig

GRASS

Let's start with the patriarch of the OSGIS world—GRASS. The
Geographic Resources Analysis Support System, or GRASS as its
commonly called, is a GIS that supports analysis, modeling, visualization, raster processing, and many other operations. It is the
"heavyweight" of the OSGIS world.

GRASS is written in C with the GUIs implemented using Python and TCL/Tk.

GRASS was originally developed by the U.S. Army Corps of Engi-
neers Construction Engineering Research Laboratories (USA-CERL)
for use in environmental research, assessments, monitoring, and
management of U.S. Department of Defense lands. The last release
was in 1992 and by 1996, USA-CERL was no longer supporting the
public in its use of GRASS. This began a transition period that in
the long run gave us the open source version of GRASS we have
today.

Today GRASS has an international team of developers and users
throughout the world, including academia, government, and con-
sulting companies. If you are interested in more of the history, visit
the GRASS home page.[2]

[2] http://grass.osgeo.org

Is GRASS a GUI program or a command-line program? The answer
is both. Seriously, though, GRASS has a GUI component, but the
real work is done by a suite of command-line programs, or *modules*,
that do everything from import data to combining grids. The GUI
side provides both a means to view your data and to perform the
many functions that GRASS provides. So in reality, you can think
of GRASS as a bit of a hybrid with the power of the command
line and the convenience of a GUI. The individual programs can be
glued together with a scripting language (shell, Perl, Python, Ruby,
your choice) to perform complex operations.

In addition to the GRASS GUIs, Quantum GIS supports viewing of
GRASS layers, giving you another option for visualizing your data.

Some of the major features of GRASS are:

• Mature and stable implementation

- Huge feature set for visualization and analysis of both raster and vector data
- Supports wide array of data formats
- Good vector digitizing tools
- Good community support from mailing lists, Wiki, and IRC
- 3D visualization
- Can be automated and scripted using common languages
- Choice of GUIs
- Good documentation
- Packages available for most supported platforms
- Raster map algebra and simulation models

What kind of user would want to use GRASS? Although it's definitely not for Clive, our casual user (he should use the QGIS-GRASS integration), it's a good choice for our advanced user Alyssa. Intermediate users will find parts of it that may be useful and worth a test drive. Only you can tell whether it's for you.

gvSIG is written in Java.

gvSIG

gvSIG is an open source project that allows you to work with a variety of vector and raster data formats, including shapefiles, PostgreSQL, MySQL, GML, GeoTIFF, ECW, JPEG, WMS, WFS, and WCS.

gvSIG provides a set of editing tools for maintaining your data and a number of geoprocessing functions.

gvSIG is multiplatform, running on Windows, Linux, and Mac OS X. Plugins can be used to extend the functionality and provide access to additional data formats.

Key features of gvSIG include:

- Good format support including web-deliverable data
- Extensible through plugins
- Editing and drawing tools
- Map layouts
- Geoprocessing tools (buffer, intersection, union, and so on)
- Raster and remote sensing tools

- 3D and animation
- Topology support

JUMP and OpenJUMP

JUMP stands for the Java Unified Mapping Platform. OpenJUMP is based on JUMP, created by Vivid Solutions,[3] with code contributions from Refractions Research.[4]

As development slowed, a group of JUMP users decided to continue development under the name OpenJUMP. This group created a "fork" of the JUMP source code and continued development apart from the JUMP development team. As a result, the two programs are similar, but different and incompatibilities have been introduced. For example, you can't use all OpenJUMP plugins in JUMP, and vice versa. In addition, there are no less than four other derivatives based on the original JUMP work at various stages in its development. These include DeeJUMP, SkyJUMP, PirolJUMP, and KOSMO.

The latest version of JUMP (1.2) released in November 2006 added support for rasters, spatial databases (PostGIS), and enhanced query capability. Most of these features were subsequently ported to OpenJUMP. While development on JUMP seems to have come to a standstill, the development of OpenJUMP continues with volunteer efforts.

While not as strong on the analysis end of the spectrum as GRASS, it does have a lot of useful features for all classes of users. Its support of GIS standards and good feature set make it attractive to a lot of folks.

While JUMP is still a viable alternative depending on your needs, let's look at some of the key features of OpenJUMP:

- Easy install
- Editing tools
- Support for PostGIS, MySQL, shapefiles, GML, and more
- WMS and WFS support

JUMP and OpenJUMP are written in Java.

[3] http://www.vividsolutions.com
[4] http://refractions.net

- Analysis functions including buffering and spatial operations
- Geometry validation—checks to see whether your features are valid
- Good set of visualization options
- Extensible framework for creating customizations
- Supports industry standards
- Multi-platform support

OSSIM

OSSIM is written in C++ and uses the Qt class library.

OSSIM stands for Open Source Software Image Map and is pronounced as "awesome." The core of OSSIM is a library that provides remote sensing, image processing, and geospatial functionality. Some of the things you can do with OSSIM include the following:

- Ortho rectification[5]

- Terrain correction

- Create mosaics from individual images

[5] *Ortho rectification* is a process where an image (photo) is registered to map coordinates (a projection) and correction is made for distortions because of terrain. The result is an image that can be used in your GIS software that will "line up" with your other data.

OSSIM supports a wide range of projections as well as a lot of data formats. As you can guess from the name, OSSIM is focused on raster processing and display rather than vector data.

You might be wondering why are we talking about a software library in the survey of desktop applications. The reason is because OSSIM also comes with a selection of command-line utilities, as well as GUI applications (ImageLinker and ossimPlanet).

ossimPlanet is a globe-style viewer that provides support for a number of formats. There are a number of reasons why you might want to use it instead of, or in addition to, Google Earth, including access to Worldwind[6] data, the ability to add your own data without having to convert it first, and its support for WMS layers. Plus, it's open source, and you can drag and drop your data right onto the globe.

[6] http://worldwind.arc.nasa.gov/index.html

Whether the OSSIM suite of programs is right for you depends on how much you play with raster data. In general, ossimPlanet is a good viewer of raster imagery available from Worldwind sources and other WMS servers around the Internet, including NASA JPL. At this point I'd say the raster processing capabilities are for the advanced user. But casual and intermediate users might find it a great tool for viewing raster data from a number of sources across the Internet. The fact that you can drag and drop your own shapefiles or Google Earth KML files right into ossimPlanet is a great advantage too. Just keep in mind that your shapefiles have to be in geographic (read latitude/longitude) coordinates in order to display them in ossimPlanet.

Some of the key features of OSSIM are:

- Impressive raster processing capabilities
- Access to a huge repository of data on the Internet through using ossimPlanet
- Support for GDAL data formats
- Runs on most platforms
- Community support via website, mailing list, and IRC

OpenEV

OpenEv allows you to view vector, raster, and WMS data sources. OpenEv has been around for a while (since 2000), and recently development on it seems to have slowed. The latest version (1.8) was released in 2004, and the project is currently in maintenance mode.[7] OpenEv is available only for Linux and Windows.

OpenEv is written in C and makes extensive use of Python.

[7] See `http://openev.sourceforge.net` for details

Should you consider using OpenEv? Frankly, there are probably better choices for you to use in visualizing your data. While there is nothing wrong with OpenEv, the facts that it's in maintenance mode and that version 2 hasn't surfaced yet makes it less attractive.

Quantum GIS

I'll try to maintain some objectivity is this section. The Quantum GIS project was founded in early 2002 with the original goal

I founded the Quantum GIS project.

306 CHAPTER 16. APPENDIX A: SURVEY OF DESKTOP GIS SOFTWARE

Quantum GIS is written in C++ and
uses the Qt class library.

of building a GIS data viewer for Linux that was fast and supported a wide range of data stores, in particular the PostGIS spatial
database. Since then, QGIS, as it's known, has grown to support
a large array of data types and runs on many platforms, including
Mac OS X, Windows, BSD, and of course Linux. The project has a
strong developer community and a world-wide user base.

QGIS provides viewing for both vector and raster data sets. Support for the majority of these is provided through the GDAL/OGR
libraries (see Section 13.2, *Using GDAL and OGR*, on page 212).
Additionally, it supports PostGIS layers (stored in a PostgreSQL
database), Spatialite, delimited text, WMS, WFS, GPS tracks, routes,
and waypoints, and GRASS layers. With QGIS you can edit PostGIS
and GRASS layers and do heads-up digitizing.

QGIS and GRASS

Recent versions of QGIS have implemented an impressive array
of geospatial analysis tools. For those tools that are not yet implemented, QGIS supports viewing, editing, and manipulation of
GRASS data through its plugin facility. This allows you to create data from the GRASS command-line and view it in QGIS. The
GRASS toolbox provided by the plugin allows you to perform a
wide array of GIS analysis and data management without leaving
QGIS.

QGIS is an application that has something for every class of user.
Casual users will find it a handy tool for visualizing data and working with your GPS data. Irving and his cohorts (that's you intermediate users) can use QGIS to create and edit data in a number of
formats. For the advanced crowd (Alyssa and friends), you can
perform analysis using the Vector and Raster tools as well as the
GRASS plugin.

Here are some of the key features of QGIS:

• Support for a wide range of both vector and raster data
• Editing capabilities

- Good set of tools for symbolizing and visualizing your data
- Map composition tools
- Good documentation
- Strong community support through forum, mailing lists, and IRC
- Extensible through plugins
- Includes a plugin for working with GPS units
- Good integration with GRASS visualization, editing, and analysis functions
- Customization using Python to write new tools and plugins

QGIS likely has a place in your visualization and editing toolbox, especially when you consider the wide range of formats supported, the integration with GRASS, and the extensibility provided by the Python bindings.

Thuban

Thuban has been around since 2002 and is developed and maintained by Intevation GMBH. Thuban provides viewing of GIS data stored in shapefiles, GeoTIFF, and PostGIS. It has projection support and can do table queries and joins.

Thuban is written in Python and uses the wxWidgets toolkit.

Thuban is a nice application with a pretty complete feature set for visualizing data. Since it doesn't provide much in the way of analysis capability, it's not going to fulfill all the needs of an advanced user. The latest release of Thuban was version 1.2.2 in September, 2009.

The key features of Thuban are:

- Lightweight viewer
- Good feature set
- Projection support
- Table queries and joins
- Support for a good range of data formats

uDig

uDig is the User Friendly Desktop Internet GIS, created by Re-

uDig is written in Java and uses the Eclipse framework.

fractions Research (`http://udig.refractions.net`). uDig provides viewing and editing for a variety of data formats, including the usual file-based layers (shapefiles and rasters), PostGIS layers, WMS, WFS, Oracle Spatial, and DB2. That should cover most of your data needs.

uDig provides a fairly complete set of viewing and editing tools. Since it supports WMS, we can pull in a wide array of free data from the Internet.

uDig is another one of those Swiss Army knife–type applications that provides a lot of features. You may not use all of them, but there is something there for pretty much everybody. If you are heavy into analysis like Alyssa, you'll find it a bit light on that end of things. Otherwise, it's an easy-to-install and easy-to-use viewer/editor.

[8] `http://udig.refractions.net/gallery`

The uDig Application Gallery[8] lists a number of custom applications illustrating the extensibility of the platform.

The key features of uDig are:

- Support for a wide range of data formats
- Easy to install
- Both viewing and editing capabilities
- Extensible
- Multi-platform support
- On-the-fly projection support

16.2   Command-Line Applications

Now it's time to take a look at some of the command-line applications you will find useful in your OSGIS ventures. With the advent of "modern" GIS software, most people want to point and click their way through life. That's good, but there is a tremendous amount of flexibility and power waiting for you with the command line. Many times you can do something on the command line in a fraction of the time you can do it with a GUI. The applications we'll look at next are definitely worthy of consideration when you start stuffing gadgets into your toolbox.

GDAL/OGR

Let's take a look at GDAL and OGR. These two are used under the hood in a large number of GIS applications, both open source and proprietary. GDAL and OGR are really libraries that provide support for a vast number of raster and vector formats. Along with the libraries are a suite of command-line tools to work with these formats.

GDAL and OGR are written in C and C++.

Raster support is provided by the GDAL library. Most popular raster formats are supported, including TIFF, PNG, JPEG2000, GRASS raster, ArcInfo grid, DEM, and ECW. Some of these formats require external libraries that are not included with GDAL (an example is ECW). If you want to have support for one of these, you will have to download and possibly build the dependent libraries and then compile GDAL. You can find a complete list of supported formats on the GDAL home page.[9]

[9] http://www.gdal.org

Vector support is provided by the OGR library. It too supports a large number of formats, including shapefiles, MapInfo (tab and mid/mif), PostGIS, GML, DGN, Oracle Spatial, and SQLite. The OGR library has similar constraints as GDAL. You may have to provide your own version of nonfree libraries needed to compile OGR (for example, Oracle Spatial).

GDAL-Supported Formats

As I said earlier, GDAL provides a number of command-line utilities for manipulating common raster formats. Operations such as warping, converting, and merging are all supported. You can also do coordinate transformation (that means changing projections) when converting between formats.

The list of supported formats in quite long. Some of the more popular raster formats supported by GDAL, and ones you are likely to encounter, are shown in the following list. We haven't listed them all—to get the full list, see the GDAL website.[10]

[10] http://www.gdal.org

- Arc/Info ASCII Grid
- Arc/Info Binary Grid (adf)
- First Generation USGS DOQ (doq)
- New Labeled USGS DOQ (doq)
- ERMapper Compressed Wavelets (ecw)
- ESRI .hdr Labelled
- ENVI .hdr Labelled Raster
- GMT Compatible netCDF
- GRASS Rasters
- TIFF/GeoTIFF (tif)
- GXF—Grid eXchange File
- Hierarchical Data Format Release 4 (HDF4)
- Hierarchical Data Format Release 5 (HDF5)
- Erdas Imagine (img)
- JPEG JFIF (jpg)
- JPEG2000 (jp2, j2k)
- MrSID
- NetCDF
- Portable Network Graphics (png)
- ArcSDE Raster
- USGS SDTS DEM (*CATD.DDF)
- USGS ASCII DEM (dem)

The list of formats may seem a bit daunting—don't worry if you don't recognize all of them; the point is to illustrate the wide range available. At last count, GDAL included support for 126 different raster formats.

GDAL Utilities

Let's take a brief look at the command-line utilities that are part of GDAL. We won't cover them all, just the ones that you're likely to find useful from the outset. For a complete list and documentation, see the GDAL website.

gdalinfo

This handy utility reports information about a raster, including, if applicable, the coordinate system, color palette, extents, and probably more than you want to know about your raster. This is a quick way to examine a raster and get some information on it without having to load it into a desktop application.

gdal_translate

This command allows you to copy a raster file and convert it to another format. You can also add coordinate system information to the output or create a subset of the image by specifying a subwindow in pixel coordinates. Another neat trick is setting a certain value in the output to nodata, making it transparent (depending of course on the software you use to view it).

gdaladdo

This utility adds overviews (commonly called *pyramids*) to a raster to improve the display speed at smaller scales. On an image with pyramids, as you zoom in, more detail will appear. One note of caution, when using gdaladdo, your original image may be modified. For a GeoTIFF the pyramids can be stored right in the original image or externally. It's a good idea to create a backup before running gdaladdo.

gdalwarp

With gdalwarp, you can "warp," in other words, transform, a raster from one coordinate system to another. This comes in handy if you want to convert a raster into a local projection to match your other data. You can also specify multiple input files to create a mosaic from a group of rasters.

gdaltindex

If you use MapServer, you'll find this utility handy for creating an index of the area covered by a group of rasters. Using gdaltindex you can create a tile index (a shapefile) that can be used with MapServer. You could also use the tile index as a coverage map to see where your rasters are located.

gdal_contour

This utility will create contour lines from a Digital Elevation

Model (DEM), which is essentially a raster composed of cells of a given size. Each cell has one attribute—an elevation. The smaller the cells, the more accurately the elevation is depicted. With gdal_contour, you specify the contour interval, and it kindly creates a shapefile with contour lines. There are other ways to do this, but gdal_contour is a quick way to make contour lines.

gdal_merge.py

If you have a bunch of adjacent images, gdal_merge.py allows you to mosaic them together to create a single seamless image. This can be handy, for example, when piecing together DRGs for a local area so you can load just the one image along with your other data.

gdal-config

This utility is used to print the configuration options and other information about the installed version of GDAL, including which raster formats are supported. If you end up getting crazy and compiling your own OSGIS software, gdal-config is almost always used during the configuration process to set things up for compiling with the GDAL/OGR libraries.

OGR-Supported Formats

While GDAL is used with rasters, OGR provides tools to manipulate vector GIS layers. OGR supports a wide variety of formats. In some cases, the OGR library can create many of the formats it supports while others are read-only. Many of the formats can also be georeferenced[11] using the library or the OGR utilities. OGR also has an impressive list of supported formats, some of which I've listed for you here. For the complete list, see the GDAL website.

[11] In this case, georeferencing means the format can contain information about the coordinate system.

- Arc/Info Binary Coverage
- Comma-Separated Value (csv)
- DWG
- DXF
- ESRI Personal GeoDatabase

- ESRI ArcSDE
- ESRI Shapefile
- GML
- GMT
- GPX
- GRASS
- KML
- Mapinfo File
- MySQL
- ODBC
- Oracle Spatial
- PostgreSQL
- SDTS
- SQLite
- U.S. Census TIGER/Line
- VRT—Virtual Datasource
- Informix DataBlade

OGR Utilities

Three utilities come with the OGR library, two of which you are likely to find very useful when working with vector data.

ogrinfo
   This utility displays information about a layer, including the coordinate system, attributes, and number of features. Given the right command-line switches, it will even print out the coordinates of every feature. You'll find ogrinfo to be very handy, especially with data you download or are handed from a stranger. It's even helpful with your own data, when you go back to it months after first creating it and can't remember anything about it.[12]

[12] Or maybe I'm the only one with this problem.

ogr2ogr
   With ogr2ogr, you can convert a vector layer from one format to another, optionally translating the coordinate system along the way. This can be especially handy when you get some new piece of data and want to get it into your favorite format or coordinate

system.

ogrtindex

Unless you're a MapServer user, you likely won't use `ogrtindex`. It creates a tile index from a group of files (for example, vectors) that can be used with MapServer.

[13] In fact, almost all of the GDAL and OGR utilities accept the `--formats` switch.

Here's a tip that you'll find useful: both the `gdalinfo` and the `ogrinfo` commands accept a `--formats` switch.[13] This is a quick way to find out which formats are supported for a given installation of GDAL/OGR. This is important because both GDAL and OGR can be compiled and distributed with support for a number of optional features. If in doubt, using `--formats` is a quick way to see whether the magic you are about to attempt is supported.

For a more comprehensive look at both the GDAL and OGR utilities, see Section 13.2, *Using GDAL and OGR*, on page 212.

Generic Mapping Tools

GMT is written in C

This next set of command-line tools can create some really impressive output. In fact, that's its whole aim—to create quality output that can be printed or included in other documents. The Generic Mapping Tools (GMT) has been around a long time. This is a testament to both its utility and acceptance by the user community. GMT was originally developed in 1988 by Paul Wessel and Walter H. F. Smith and is currently hosted at the University of Hawaii.

GMT allows you to create cartographic-quality maps from the command line. This sounds simple, but in fact it has quite sophisticated features including base map creation, plotting x-y values, lines, and polygons, coordinate transformations, gridding, contouring, and 3D illuminated surfaces.

Now I know you are probably thinking that GMT doesn't exactly fit your idea of a desktop GIS application. In fact, it can be a valuable addition to your toolkit. For an introduction to GMT and its capabilities, see Section 13.1, *GMT*, on page 199.

16.3   *Other Tools*

As I said earlier, there is a huge selection of OSGIS applications to choose from. It's also impossible to discuss each of these in detail. Our survey included some of the major applications available today—and those that together in some combination form a productive toolkit.

Our survey should have got you started thinking about some of the tools available. Now it's up to you to carry on the survey if you see fit. To get you started, here are some links to lists of OSGIS software and tools that you may want to peruse:

- `http://www.osgeo.org`
- `http://www.freegis.org`
- `http://opensourcegis.org`
- `http://en.wikipedia.org/wiki/List_of_GIS_software#Open_source_software`
- `http://maptools.org`

For a good but somewhat dated survey that includes both open source desktop and web mapping tools, see "The State of Open Source GIS" by Paul Ramsey.[14]

[14] `http://2007.foss4g.org/presentations/viewattachment.php?attachment_id=8`

17

Appendix B: Installing Software

In this appendix, you will find brief information on installing most of the applications we have discussed. As always, it helps to read the installation instructions provided with the software. The following information is of the quick-start variety and will help you get up and running.

We provide information for each platform, assuming of course that the application is supported on each.

17.1 GRASS

The good news is you can get a binary distribution for most major platforms. And if you can't, GRASS builds quite readily on most platforms. In this section, we'll look at the options for each of the major platforms so you can get up and running quickly. All binary and source packages are available from the GRASS website.[1]

[1] http://grass.osgeo.org

Linux

Binaries are readily available for Linux, including a generic `tar.gz` package. Typically, you will find binaries for the following distributions:

- Generic GNU/Linux
- Debian

- Mandriva
- OpenSuSE
- Red Hat Enterprise Linux (RHEL, CentOS, Fedora, and Scientific Linux)
- Ubuntu

To install, just use the package management tool(s) provided with your distribution (for example, rpm, apt, yum). The package for Generic GNU/Linux is easily installed using the provided installation script for your distribution. For all packages, GRASS depends on a number of supporting libraries that you will also have to install using your package management system.

In the event you can't find a package for your distribution or you just want to live on the edge, you will have to compile GRASS from source. There is a complete "Compiling source code" manual available on the website. Compiling from source is not difficult and may be your best option if you want to stay current with new developments.

Unix

If you are using Solaris, Irix, HP-UX, DEC-Alpha, AIX, or one of the BSD variants, compiling from source is your only option. GRASS should build on POSIX systems using the GNU C compiler.

OS X

For Mac OS X a binary distribution is available from the GRASS website as a disk image (dmg). Currently, a number of other frameworks are required along with the GRASS image. These are also available from the link on the website.

Installation is standard Mac fare—open the disk image, and install the package. Make sure you get all the required frameworks installed beforehand as described on the website.

Windows

On Windows you have several choices:

- Stand-alone installer for XP through Windows 7
- The OSGeo4W installer
- Cygwin

See the website for information on which is best for you.[2]   Most
users will want either the stand-alone or OSGeo4W installer.

[2] http://grass.osgeo.org/
grass64/binary/mswindows

17.2   OpenJUMP

Installing OpenJUMP consists of the following:

1. Downloading the installer (Windows) or the distribution ZIP
   file
2. Unzipping the distribution as needed
3. Running the installer or start-up script in the `bin` directory

You may need additional Java libraries to support some of the plugins. Your best bet is to refer to the install instructions on the Open-
JUMP website[3] after unzipping the distribution for the latest information on getting things up and running.

[3] http://sourceforge.net/apps/
mediawiki/jump-pilot/index.
php?title=Installation

17.3   Quantum GIS

For QGIS, you will find packages for the major platforms, including
Linux, Mac OS X, and Windows on the download site.[4] To install
on each platform, do the following:

[4] http://download.qgis.org

- *Mac OS X*: Install the needed frameworks and then use the installer package to install QGIS.
- *Windows*: Choose the installer you want to use and follow the
  instructions to install QGIS.
- *Linux*: If you find a package (`rpm`, `deb`) for your Linux distribution, just download and install it. Some Linux distributions
  include QGIS. Check the QGIS download site for information
  on the various distributions. If a package is not available, you
  will have to compile QGIS and its dependencies. Information on
  building from source is available on the QGIS website.[5]

[5] http://qgis.org

QGIS depends on a number of other OSGIS packages. When installing on Windows, these are bundled for you in the package. On Linux and other *nix platforms, you will need to install the dependencies prior to installing or building QGIS. For distributions that have a package manager, this is usually not a difficult task. It's best to look around before building dependencies from scratch to see whether packages or builds for your operating system are available. The Mac OS X version requires the installation of a couple of frameworks prior to installing QGIS. These are documented on the QGIS download website.

17.4 uDig

Installing uDig is pretty easy if you are running Windows, Linux, or Mac OS X. The uDig download site[6] provides binary releases for the three aforementioned platforms.

[6] http://udig.refractions.net/download

Installation is fairly simple for every platform; just download the package for your platform and unzip it and run it.

Currently, the Windows and Linux binary distributions include a Java runtime and are ready to run. On Mac OS X, Java is already installed, and you are good to go. Since all the dependencies needed for uDig are packaged with it, you don't need to download and install anything else. This makes it quick and easy to get going.

17.5 GMT

As with most OS GIS applications, you have choices when it comes to installing GMT. Many Linux distributions include GMT. Check your package management tool to see what's available.

The GMT website has detailed instructions for installing on Unix, Linux, Windows, and Mac OS X.[7]

[7] http://gmt.soest.hawaii.edu

If you are a Windows user, your options are a bit more limited. The GMT folks suggest installing Cygwin and then building GMT as you would for Unix. Again, see the website for your options.

17.6 GDAL/OGR

- *Linux*: Binaries for GDAL and OGR are available from the GDAL website.[8] For Linux, check the current packages, and install using your package manager. If you don't find current binaries for your Linux distribution, then you will have to build from source or install FWTools (Section 17.7, FWTools).

  [8] http://trac.osgeo.org/gdal/wiki/DownloadingGdalBinaries

- *Mac OS X*: Binaries frameworks are available from the Kyng Chaos website.[9] Just download and install the package per the instructions found in the `ReadMe.rtf`.

  [9] http://www.kynchaos.com

- *Windows*: Links to binaries and plugins for Windows can be found on the OSGeo download site.

For Linux and Windows, you may find that FWTools is a better way to go—you get GDAL/OGR as well as a bunch of other handy applications.

17.7 FWTools

FWTools[10] is an open source GIS binary kit for Linux and Windows. It includes GDAL/OGR, the projections library PROJ4, MapServer, and Python, among others. It is a complete runtime environment with no external dependencies. Install it, and you are ready to use all the included applications and utilities.

[10] http://fwtools.maptools.org

To install FWTools, do the following:

- *Linux*: Download and untar the distribution, change to the `FWTools` directory, and execute the `install.sh` script. Once it's complete, you'll need to add the `bin_safe` directory to your path.
- *Windows*: Download the installer and run it, and then follow the instructions to complete the installation. The installation creates a shortcut to start an FWTools shell from which you can use the applications.

Note that FWTools may not contain the latest versions of the included software. If not, you'll have to install the applications you want individually to get access to the latest features and tools.

18

Appendix C: GRASS Basics

Once you have GRASS installed, setting up GRASS and creating your locations is key to getting off the ground. This appendix will guide you through creating a location using both QGIS and the GRASS shell. From there you'll get a basic introduction to working with GRASS GIS. Once you've mastered the basics of GRASS, you might be interested in diving deeper with *Open Source GIS: A GRASS GIS Approach*.[1]

[1] Markus Neteler and Helena Mitasova, 2008, *Open Source GIS: A GRASS GIS Approach*. Third Edition. The International Series in Engineering and Computer Science: Volume 773. 406 pages, 80 illus., Springer, New York

18.1 Location, Location, Location

Why are we repeating ourselves? Because it's real important. One of the key concepts in using GRASS is that of a *location*. Remember the old axiom about location used in both real estate and business? Well, the same holds true for GRASS. Once you get the concept down, you will have mastered what many have stumbled on.

Quick Start

We are going to "cheat" to get started with GRASS data by using the QGIS/GRASS plugin to create a new location and mapset. We'll then go back and look at alternative ways to do the same. This method makes it very easy to create a new location. We simply add one or more layers to QGIS that encompass the geographic region of interest and then use the plugin to do the dirty work. Before we do that, though, we need a GRASS database to get started with.

A GRASS database is simply a directory where we will store our GRASS locations and mapsets, along with the data. You can have as many GRASS databases as you desire. For this example, we'll use a directory in our home directory named grassdata. The QGIS/-GRASS plugin can create this directory as part of defining our new location and mapset.

Creating the Location using QGIS

If you don't already have the sample data, you can get it from http://geospatialdesktop.com.

Let's start by running QGIS and loading the data that covers the area of interest. We'll use the sample data here and load just the world borders layer.

Figure 18.1: GRASS database selection/creation dialog box

We are now ready to create the database, location, and mapset.

Choose GRASS from the Plugins menu and choose New Mapset. Here
you can choose an existing database directory or create a new one.
The dialog box provides some visual cues as to how you should proceed. In Figure 18.1, on the facing page, you can see the database
selection dialog box with values for our new database plumbed in
(to create a new database, just create a new folder using the browse
button to the right of the Database field).

Clicking the Next button brings us to the *New Mapset* screen. We
can choose to create a new mapset in an existing location or use it
to create a new location, which is what we want to do. To create
a location, click the *Create new location* radio button and provide a
name, in this case world_lat_lon.

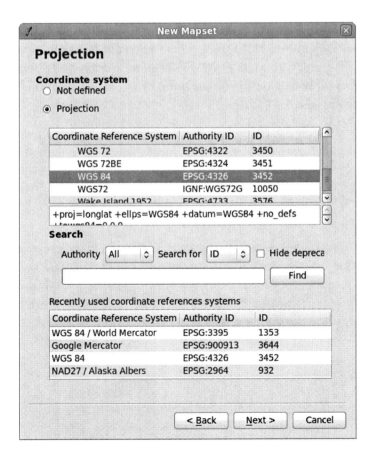

Figure 18.2: Choosing the WGS 84
projection for the GRASS location.

Clicking Next opens the projection page where we set the projection for our location. While the dialog will let you get away without setting a projection it's not a good idea. Our world_lat_lon layer is WGS 84 geographic so let's choose it from the list of Geographic Coordinate Systems as shown in Figure 18.2,on the previous page.

Moving to the next screen (by clicking the Next button) brings us to *Default GRASS Region.* Here you can customize the region (think extents) for the location. By default, it will be set to some sane value but we still need to check by comparing with the extents we want our mapset to cover. To set the region to the extents of the data loaded in QGIS click the *Set current QGIS extent* button and check the values. You can make adjustments by manually entering the N, S, E, and W values. Note that this screen also allows you to set the extent by choosing countries or regions from a drop-down list. This can be useful if you want to create a new location but don't have any layers loaded in QGIS. In this case you need to specify the region, as well as the projection information since it won't be automatically populated for you.

In Figure 18.3, on the facing page, you can see the region screen with the extents set to cover the entire planet. Notice the extents are shown graphically in red on the world map.

Keep in mind that we have defined a default region. The region can be changed for each mapset or for the entire location later.

Clicking the Next button brings us to the final screen in the process. Here we give the mapset a name. As explained in the dialog box, a mapset is simply a collection of maps (think layers here) used by a single user. You can name the mapset whatever you like—we used "work". In a multiuser environment, a user ID might be a good choice. Enter a name, and click Next. This brings up a summary of what we just did, telling us that we have created a database in our chosen directory, a named location, and a mapset.

Click Finish, and we are done. You now have a new location and mapset ready for your use. The plugin automatically opens it and

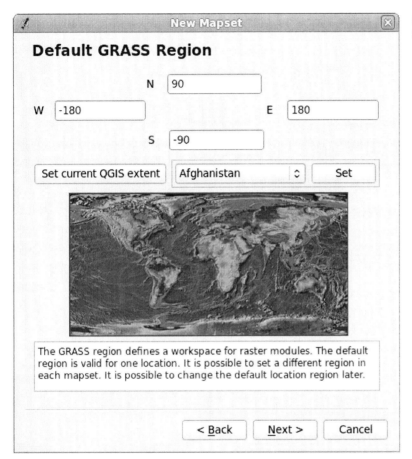

Figure 18.3: Default region settings for a WGS 84 world location

sets it as the current location. You should be able to see the region delineated by a red rectangle on the map canvas.

Using the QGIS/GRASS plugin allowed us to create the new location without knowing much at all about GRASS. Let's change that now.

Creating a Location with GRASS

You are probably noticing about now that we haven't run GRASS yet. So far everything we have done has been through QGIS using the GRASS plugin. Now it's time to show you how to start GRASS and create a location. Starting GRASS depends on your platform:

- *Linux*: From a terminal window, run the GRASS binary using `grass` combined with the version number. So to start GRASS 6.4, you would use `grass64` from the command line or the menu.
- *Mac OS X*: Start by double-clicking the GRASS icon in your `Applications` folder.
- *Windows*: Depending on which installation method you chose, start your Cygwin shell, and start GRASS by running `grass64` or start it from the menu.

On Linux GRASS remembers and starts up in the last-used mode when you start it from the command line

When you start GRASS, you are presented with either a text- based form or a GUI dialog box requesting the location and mapset you want to use for the session. Which you see depends on how you started GRASS. At present GRASS provides four user interfaces:

- Text based
- The "old" GUI based on Tcl/Tk
- The "old-old" GUI based on Tcl/Tk
- The new GUI built with wxPython

If you are a new GRASS user I suggest you use the new Python based interface. You can start this from the GRASS menu or from a command line using:

```
grass64 -wx
```

For the purpose of this example, we will use the wxPython GUI to create a new location. You could do the same with any of the GRASS interfaces, including the text mode.

In Figure 18.4, on the next page, you can see the start-up screen for GRASS 6.4.x. We have already filled in the data directory (grass-data) and are ready to create a new location using the Location wizard. If you don't enter the data directory location, the wizard will let you do it on the next screen.

If you don't have a directory for storing your GRASS locations in, you can create it before you start GRASS, or just use the Browse

Figure 18.4: GRASS start-up form

button to bring up the directory dialog and create a new folder. You can store locations in any directory, but it's best to establish a structure for your data to keep it apart from other non-GRASS files and directories on your system.

The other two things we need to decide on up front are a name for the new location and a title. The name should be descriptive so you can remember at a glance what it represents. I typically include the name of the projection in the location name. As we did in the previous example (Section 18.1 *Creating the Location using QGIS*, on page 324) we'll name the location world_lat_lon and title it something like "WGS 84 world location".

Figure 18.5: GRASS location parameters

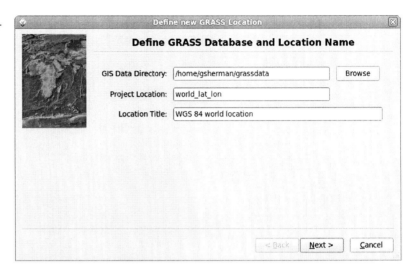

In Figure 18.5, we have entered the database directory, name, and title for our new location. When we click next we get to select the projection for our new location by:

• Selecting from a list
• Selecting an EPSG code
• Reading the projection from a georeferenced file
• Reading the projection from a WKT or PRJ projection file
• Specifying proj4 parameters
• Creating an X, Y projection (non-georeferenced)

This gives you a lot of flexibility for setting the projection. Since we are making a location that spans the world, let's use the world_lat_lon shapefile to set the projection by clicking on the appropriate option to read from a georeferenced file. Clicking Next gives us a summary of what we did, including the database location, location name, title, and the projection. Clicking Finish creates the new location and pops up a dialog asking if we want to set the default extents and resolution. Since we set the projection from a shapefile the extents are already filled in—click Yes and look at the settings and then click the *Set region* button.

That's it—we have created a new location and it is now available from the startup screen. You can see our location now has one mapset—PERMANENT, which is for shared layers. We need to create a mapset for us to use for loading our data and eventually doing some editing. To create a mapset, click on the *Create mapset* button and enter a name for it. We've used "work" but you can use any reasonable name as we mentioned in a previous section. Once you have created the mapset, you can highlight it in the startup screen and click the *Start GRASS* button to bring up the GRASS GUI. Figure 18.6 shows the startup screen with our new location and mapset selected and ready to launch.

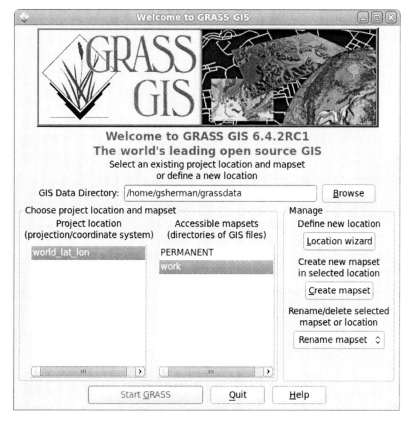

Figure 18.6: GRASS location and mapset created and ready to use.

In this section, we saw a couple of ways to create a new location. We didn't look at them all but just enough to get us started. Of all the methods available to us, using QGIS or the Georeferenced File method in the GRASS GUI are probably the quickest and easiest. If you are familiar with projections and EPSG codes, you may find the other methods just as easy.

18.2   Getting Some Data

A GRASS location and mapset is nice, but you obviously need some data to work with. GRASS uses its own format for storing data. Unlike the OSGIS viewers we have discussed (Thuban, uDig, QGIS) that read a number of formats, GRASS prefers its own format. That's not to say you can't use external data with GRASS because you can. To get full advantage of the capabilities, importing the data is required.

This section will show you several ways to get some data into GRASS so you can work with it, including the following:

- Import the data using `v.in.ogr` and `r.in.gdal` GRASS commands.
- Use the GRASS plugin in QGIS to import a loaded vector layer.
- Use `v.external` to make an external layer (for example a shapefile) accessible to GRASS.

Using the Command Line

Let's start by importing some vector data into GRASS using `v.in.ogr`. This command allows you to import data sources supported by the OGR library (see Section 16.2, *OGR Utilities*, on page 313 for information on supported formats). This means we can use it to import a shapefile, among others. The `world_borders` shape is stored in the `desktop_gis_data` folder (if you are following along, your location may vary). We will use it to import into GRASS.

Let's look at part of the syntax help for `v.in.ogr`. If you want to get the full picture, use `g.manual v.in.ogr`:

```
GRASS 6.4.2RC1 (world_lat_lon):~ > v.in.ogr help

Description:
 Convert OGR vector layers to GRASS vector map.

Keywords:
 vector, import

Usage:
 v.in.ogr [-lfcztorew] [dsn=string] [output=name]
   [layer=string[,string,...]]
   [spatial=xmin,ymin,xmax,ymax[,xmin,ymin,xmax,ymax,...]] [where=sql_query]
   [min_area=value] [type=string[,string,...]] [snap=value]
   [location=string] [cnames=string[,string,...]] [--overwrite]
   [--verbose] [--quiet]

Flags:
  -l   List available layers in data source and exit
  -f   List supported formats and exit
  -c   Do not clean polygons (not recommended)
  -z   Create 3D output
  -t   Do not create attribute table
  -o   Override dataset projection (use location's projection)
  -r   Limit import to the current region
  -e   Extend location extents based on new dataset
  -w   Change column names to lowercase characters
  --o  Allow output files to overwrite existing files
  --v  Verbose module output
  --q  Quiet module output

Parameters:
      dsn   OGR datasource name
              Examples:
ESRI Shapefile: directory containing shapefiles
MapInfo File: directory containing mapinfo files
   output   Name for output vector map
    layer   OGR layer name. If not given, all available layers are imported
              Examples:
ESRI Shapefile: shapefile name
MapInfo File: mapinfo file name
  spatial   Import subregion only
              Format: xmin,ymin,xmax,ymax - usually W,S,E,N
    where   WHERE conditions of SQL statement without 'where' keyword
              Example: income < 1000 and inhab >= 10000
 min_area   Minimum size of area to be imported (square units)
              Smaller areas and islands are ignored. Should be greater than snap^2
              default: 0.0001
     type   Optionally change default input type
              options: point,line,boundary,centroid
              default:
```

```
            point: import area centroids as points
            line: import area boundaries as lines
            boundary: import lines as area boundaries
            centroid: import points as centroids
   snap     Snapping threshold for boundaries
            '-1' for no snap
            default: -1
location    Name for new location to create
  cnames    List of column names to be used instead of original names, first
            is used for category column
```

If you happen to invoke a GRASS command without any parameters, a GUI window will pop up, allowing you to set the options and execute the command. To import our layer, we'll use the command line rather than the GUI but feel free to experiment.

There are a lot of optional flags and parameters we can use with v.in.ogr. For our example, we are going to go the simple route and see whether it gives us the expected results. To import the layer, use the following:

```
v.in.ogr dsn=/home/gsherman/desktop_gis_data output=world_borders \
  layer=world_borders
```

The dsn is the data source name and, in the case of OGR vectors, refers to the directory where the layers are stored. In this case, that's the directory where the world_borders shapefile resides: /home/gsherman/desktop_gis_data. If you don't specify a layer name using the layer parameter, all files in the dsn directory will be imported into separate layers in a single map. This obviously can be pretty handy for bringing in a lot of layers all at once—but only if having them lumped into a single map works for you.

> **GRASS Commands**
>
> ───────────────────
>
> GRASS commands are arranged by function. For example, vec-
> tor commands begin with v., raster commands begin with r.,
> database commands with d., and general commands with g..
> For a complete list of these commands and their function, refer
> to the online GRASS manual that is installed with GRASS. You
> can easily access the manual using the g.manual -i command
> from the GRASS shell.

Importing the layer will take a while. GRASS does a number of
things when importing a layer, including building topology as appropriate. Part of the output from the import process is shown next.
We didn't include the entire output because it gets quite long and
includes detailed information about the import process and building topology.

```
GRASS 6.4.2RC1 (world_lat_lon):~ > v.in.ogr dsn=/home/gsherman/desktop_gis_data \
  output=world_borders layer=world_borders
Projection of input dataset and current location appear to match
Layer: world_borders
Counting polygons for 3784 features...
Importing map 3784 features...
 100%
--------------------------------------------------
Building topology for vector map <world_borders_tmp>...
```

> ## What is Topology?
>
> You probably noticed the word *topology* has cropped up a couple of times now. In simple terms, topology is the relationship between spatial features. For example, a polygon boundary consists of lines. Adjacent polygons share common boundaries. Data formats that are topological maintain this relationship when creating and editing data. The common boundaries are stored only once.
>
> In a nontopological format, the common boundaries are duplicated, one for each polygon.
>
> Apart from the storage difference, a topological GIS such as GRASS maintains the spatial relationships as you edit data. If you move a line, the polygon boundary or boundaries are adjusted accordingly. This provides consistency in your data and is essential when performing many geoprocessing tasks.

If the import is successful, you should be returned to the GRASS prompt without any error messages. You can quickly confirm that we now have a world_borders layer using the following:

```
GRASS 6.4.2RC1 (world_lat_lon):~ > g.list vect
----------------------------------------------
vector files available in mapset <work>:
world_borders
----------------------------------------------
```

You can also set the region using the GUI.

Now let's bring in a raster using the r.in.gdal command. This command uses the GDAL library and therefore can import a wide range of formats. For this example, we will import the NASA world mosaic into GRASS. Before we import, we need to make sure that we set the region to cover the area of interest—in this case the entire world, and set the resolution to appropriate values. Since we have imported the world_borders vector layer, we can use it to set the region:

The rows and columns used to set the resolution match those of the world_mosaic.tif. You can get this information using gdalinfo.

```
GRASS 6.4.2RC1 (world_lat_lon):~ > g.region vect=world_borders \
  rows=8192 cols=4096
```

We are now ready to give the import a try. The syntax of r.in.gdal is as follows:

```
GRASS 6.4.2RC1 (world_lat_lon):~ > r.in.gdal help

Description:
 Import GDAL supported raster file into a binary raster map layer.

Keywords:
 raster, import

Usage:
 r.in.gdal [-oeflk] [input=name] [output=name] [band=value]
   [memory=value] [target=name] [title="phrase"] [location=name]
   [--overwrite] [--verbose] [--quiet]

Flags:
  -o   Override projection (use location's projection)
  -e   Extend location extents based on new dataset
  -f   List supported formats and exit
  -l   Force Lat/Lon maps to fit into geographic coordinates (90N,S; 180E,W)
  -k   Keep band numbers instead of using band color names
 --o   Allow output files to overwrite existing files
 --v   Verbose module output
 --q   Quiet module output

Parameters:
     input   Raster file to be imported
    output   Name for output raster map
      band   Band to select (default is all bands)
    memory   Cache size (MiB)
    target   Name of location to read projection from for GCPs transformation
     title   Title for resultant raster map
  location   Name for new location to create
```

The options here are fewer than with v.in.ogr. To import the raster,
use the following:

```
GRASS 6.4.2RC1 (world_lat_lon):~ > r.in.gdal ev11612_land_ocean_ice_8192.tif \
  output=nasa_world_mosaic
ERROR: Projection of dataset does not appear to match current location.

        Location PROJ_INFO is:
        name: Lat/Lon
        proj: ll
        datum: wgs84
        ellps: wgs84
        no_defs: defined

        Import dataset PROJ_INFO is:
        cellhd.proj = 0 (unreferenced/unknown)

        You can use the -o flag to r.in.gdal to override this check and use
```

```
the location definition for the dataset.
Consider generating a new location from the input dataset using the
'location' parameter.
```

Oops—what happened? GRASS doesn't like our raster because it thinks it may be in a different projection than our WGS84 location. We can override this using the -o switch if we're sure the projection is correct:

```
GRASS 6.4.2RC1 (world_lat_lon):~ > r.in.gdal -o ev11612_land_ocean_ice_8192.tif \
  output=nasa_world_mosaic
WARNING: Over-riding projection check
WARNING: G_set_window(): Illegal latitude for North
```

Now what? It appears that GRASS isn't happy with the raster's northern latitude value(s). Let's use gdalinfo to examine the raster so we can determine what's going on:

```
GRASS 6.4.2RC1 (world_lat_lon):~ > gdalinfo -nomd \
  desktop_gis_data/ev11612_land_ocean_ice_8192.tif
Driver: GTiff/GeoTIFF
Files: desktop_gis_data/ev11612_land_ocean_ice_8192.tif
       desktop_gis_data/ev11612_land_ocean_ice_8192.tfw
Size is 8192, 4096
Coordinate System is ''
Origin = (-180.021972650000009,90.021972649999995)
Pixel Size = (0.043945300000000,-0.043945300000000)
Corner Coordinates:
Upper Left  (-180.0219727,  90.0219726)
Lower Left  (-180.0219727, -89.9779762)
Upper Right ( 179.9779249,  90.0219726)
Lower Right ( 179.9779249, -89.9779762)
Center      (  -0.0220239,   0.0219982)
Band 1 Block=8192x1 Type=Byte, ColorInterp=Red
Band 2 Block=8192x1 Type=Byte, ColorInterp=Green
Band 3 Block=8192x1 Type=Byte, ColorInterp=Blue
Band 4 Block=8192x1 Type=Byte, ColorInterp=Undefined
```

This provides us with a lot of detail information about the raster. For now, we are interested in the coordinates. Notice that our image extends beyond the bound of the world (at least as we defined it in our GRASS location). The image appears to be shifted to the upper left. This is because a world file specifies the position of the *center* of the upper-left pixel. You can see that the shift is exactly half the pixel size of 0.04394530. So, how do we solve this problem so we can complete the import? The answer is to simply translate the bounding rectangle so it is correct. To do that, we will use the gdal_translate command:

```
GRASS 6.4.2RC1 (world_lat_lon):~ > gdal_translate -of GTiff \
  -a_ullr -180 90 180 -90 -co "COMPRESS=LZW" -a_srs EPSG:4326 \
  ./desktop_gis_data/evl1612_land_ocean_ice_8192.tif world_mosaic.tif
Input file size is 8192, 4096
0...10...20...30...40...50...60...70...80...90...100 - done.
```

Using gdal_translate we translated the raster and created a new one with the proper bounding rectangle (-a_ullr -180 90 180 -90). We didn't change the data; all we did was remove the shift because of the way world files are designed. During the process, we specified the image should be compressed using LZW (-co "COMPRESS=LZW") and also assigned the WGS84 projection to it (-a_srs EPGS:4326). The new image is a true GeoTIFF. It doesn't have a world file but has the projection information encoded in the file itself. Using gdalinfo on the new world_mosaic.tif yields the following:

```
GRASS 6.4.2RC1 (world_lat_lon):~ > gdalinfo desktop_gis_data/world_mosaic.tif
Driver: GTiff/GeoTIFF
Files: desktop_gis_data/world_mosaic.tif
Size is 8192, 4096
Coordinate System is:
GEOGCS["WGS 84",
    DATUM["WGS_1984",
        SPHEROID["WGS 84",6378137,298.257223563,
            AUTHORITY["EPSG","7030"]],
        AUTHORITY["EPSG","6326"]],
    PRIMEM["Greenwich",0],
    UNIT["degree",0.0174532925199433],
    AUTHORITY["EPSG","4326"]]
Origin = (-180.000000000000000,90.000000000000000)
Pixel Size = (0.043945312500000,-0.043945312500000)
Metadata:
  TIFFTAG_XRESOLUTION=72
  TIFFTAG_YRESOLUTION=72
  TIFFTAG_RESOLUTIONUNIT=2 (pixels/inch)
  AREA_OR_POINT=Area
Image Structure Metadata:
  COMPRESSION=LZW
  INTERLEAVE=BAND
Corner Coordinates:
Upper Left  (-180.0000000,  90.0000000) (180d 0' 0.00"W, 90d 0' 0.00"N)
Lower Left  (-180.0000000, -90.0000000) (180d 0' 0.00"W, 90d 0' 0.00"S)
```

```
Upper Right ( 180.0000000,  90.0000000) (180d 0' 0.00"E, 90d 0' 0.00"N)
Lower Right ( 180.0000000, -90.0000000) (180d 0' 0.00"E, 90d 0' 0.00"S)
Center      (   0.0000000,   0.0000000) (  0d 0' 0.01"E,  0d 0' 0.01"N)
Band 1 Block=8192x1 Type=Byte, ColorInterp=Red
  Mask Flags: PER_DATASET ALPHA
Band 2 Block=8192x1 Type=Byte, ColorInterp=Green
  Mask Flags: PER_DATASET ALPHA
Band 3 Block=8192x1 Type=Byte, ColorInterp=Blue
  Mask Flags: PER_DATASET ALPHA
Band 4 Block=8192x1 Type=Byte, ColorInterp=Alpha
```

The important things to note in the output of gdalinfo are as follows:

- The coordinate system WGS84 has been assigned and encoded in the file.
- The pixel size hasn't changed (apart from some minor rounding).
- The bounding coordinates are now correct.

At last we can import the image into GRASS. Now that the image has been tortured into submission, we don't have to supply any flags to the r.in.gdal command:

```
GRASS 6.4.2RC1 (world_lat_lon):~ > r.in.gdal input=world_mosaic.tif \
  output=world_mosaic
Projection of input dataset and current location appear to match
 100%
r.in.gdal complete. Raster map <world_mosaic.red> created.
 100%
r.in.gdal complete. Raster map <world_mosaic.green> created.
 100%
r.in.gdal complete. Raster map <world_mosaic.blue> created.
 100%
r.in.gdal complete. Raster map <world_mosaic.alpha> created.
```

Success! We stumbled across a bit of a problem initially, but this served to illustrate some problem-solving tools and techniques at our disposal. You should note that you won't encounter the offset issue with every raster you import. If you are importing a GeoTIFF that contains coordinate information, odds are it will be processed just fine. If not, you are now prepared to deal with the problem.

Compositing combines the red, green, and blue bands to create the full color image.

The last thing we want to do is color-composite the image for display purposes. The import processed each band of the image separately, creating separate outputs for red, green, blue, and the alpha channel. We can combine these together so the image looks like we expect using r.composite:

```
GRASS 6.4.2RC1 (world_lat_lon):~ > r.composite red=world_mosaic.red \
  green=world_mosaic.green  blue=world_mosaic.blue output=world_mosaic
Creating color table for output raster map...
 100%
Writing raster map <world_mosaic>...
 100%
r.composite complete. Raster map <world_mosaic> created.
```

In Figure 18.7, you can see the completed world mosaic composite displayed in GRASS.

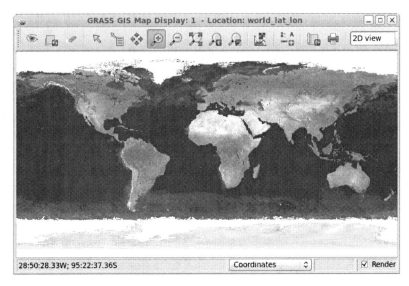

Figure 18.7: World mosaic in GRASS

Importing with QGIS

Now let's look at using QGIS to import a layer into GRASS. After QGIS is fired up, open the mapset we want to use for storing our data. To open a mapset, choose Open mapset from the GRASS menu under the main Plugins menu. Make the appropriate selections for location and mapset, and press OK. Nothing much happens when you do this, other than the GRASS toolbox icon will now be enabled.

To import a shapefile into GRASS, you have a choice of loading it into QGIS using the `Add a Vector Layer` menu or toolbar icon and then importing or going directly to the `v.in.ogr` tool in the GRASS toolbox. The difference between the tools is minimal—in one you've already loaded the layer and in the other you have to browse for it. We'll assume we've added the `cities` layer to QGIS and now want to import it using `v.in.ogr.qgis`.

With the `cities` layer loaded into QGIS, open the GRASS toolbox by clicking the tool in the GRASS toolbar or by choosing `Open GRASS tools` from the GRASS plugin menu. The toolbox contains a wealth of GRASS tools and functions we can run from within QGIS. Click on the *Modules List* tab and scroll down until you find the `v.in.ogr.qgis` tool—or you can type "ogr" in the Filter box to make it quicker to find. This tool will give us the opportunity to import any loaded vector layer into GRASS, assuming it's an OGR or PostGIS layer.

The `v.in.ogr.qgis` tool actually uses `v.in.ogr` to do the import. It just provides a front-end that allows you to select the QGIS layer you want to import.

Once we click the import tool, a new tab page is opened, and a drop-down box of all the eligible layers is available. We just pick the one we want to import from the list and then fill in a name for the GRASS layer (output vector map). There are some advanced options that we could explore but we don't need them to import this layer. Clicking the Run button starts the import process. The output from the command will be displayed as the import proceeds. If all goes well, a "Successfully finished" message will be displayed at the end of the output. If you scroll back through the output window, you will find that the command used to do the import is `v.in.ogr`. The GRASS plugin in QGIS just provided a convenient front end for importing the layer. We could easily have accomplished the same thing using the GRASS command line or GUI.

If you click the `Add GRASS vector layer` tool you will see that our GRASS mapset now contains both the `world_borders` layer and the `cities` layer. We can use the same method to import rasters into GRASS, using the `r.in.gdal` tool in the toolbox. If we encounter difficulties with the raster as we did the NASA world mosaic, some prep work using `gdalinfo` and `gdal_translate` may be required.

Importing External Data

The last means of getting vector data into GRASS involves importing external data. Actually, it's not really an import but more of a link. When working with external data, you have to be aware that a number of GRASS tools and operations won't work. But it is a handy way to display data without converting to a GRASS layer. You can link external data using the command line or the GRASS GUI File→Link external formats menu. For information on which OGR formats can be linked, see the manual for v.external.

We'll illustrate using the command line. The syntax for v.external is almost exactly like the simplest form of v.in.ogr:

```
v.external dsn=string [output=name] [layer=string] [--overwrite]
```

To link the world_borders shapefile, we would use the following, being sure to specify the full path in the dsn:

```
v.external dsn=/home/gsherman/desktop_gis_data output=world_borders_external \
  layer=world_borders
```

We can now use the layer in GRASS—just remember it is read-only and may yield incorrect results when used in some GRASS operations. This is because an external layer is fundamentally different from a true GRASS layer. For example, an external layer doesn't have true topology but a pseudo-topology is created to allow it to act like a complete GRASS layer. Again, if you want to use a layer for operations other than display, it's best to import it into GRASS using one of the methods we have discussed in this chapter.

18.3   *Working with Data*

Now that we know how to get data into GRASS, let's work with it a bit and learn how to use the display manager. As you know by now, GRASS has a several start-up modes—GUI and text. GRASS remembers which mode you used last and will attempt to start up using the same the next time around. Not only are there two modes, there are also three GUIs to choose from. By default if you start GRASS using grass64 -wx, you get the newest display manager based on wxPython. The two other GUIs are based on Tcl/Tk: `gis.m` and the much older `d.m`. For information on which switch to use on the command line to start your GUI of choice, use `grass64 -help` to get a list of all the options.

Which GRASS GUI Should You Use?

Since GRASS currently has three GUI interfaces you have to make a choice between:

- `g.gui wxPython`
- `gis.m`
- `d.m`

Unless you have a compelling reason to use one of the older GUIs, I would stick with the newest–wxPython.

Let's start up the GRASS wxPython GUI so we can explore some data using our world_lat_lon location:

Depending on your operating system and GRASS installation you may have a menu option to start the wxPython GUI.

```
grass64 -wx
```

Once the GUI initializes we have a *Layer Manager* and a *Map Display* window. Everything starts with the *Layer Manager*. Its toolbars are organized by function, in particular the raster and vector functions are those we are interested in at the moment. Like other GIS applications, GRASS has tooltip text for each button on the toolbar(s). Hover the mouse to learn what each tool is for, or consult the manual. The button to add a vector layer looks like some line segments in the shape of a "V" with a plus sign below it. Clicking it opens a dialog that allows us to choose a layer to add and set a number of options.

On the *Required* tab you can click the drop-down box to get a list of vector layers in the mapset. Once we choose `world_borders` we have a bunch of display options we can set, including:

- Geometry
- Category numbers
- Topology information

- Line directions

By default, geometry is checked and we'll go with that for now.

On the *Selection* tab we can choose the feature types we want to display:

- Point
- Line
- Boundary
- Centroid
- Area
- Face

By default all of them are checked. Remember you can store multiple layers within a single map so some or all of these feature types could be present. For the world_borders layer we are interested in the polygons that represent each country so we can clear all the checkboxes except area.

The Layer number drop-down box allows us to choose which layer we want to display. By default all layers are displayed. If you click the drop-down you'll see we only have one layer in world_borders, layer 1. There are some other things we can do on the *Selection* tab but we won't worry about those for now.

The *Colors* tab allows us to style the display of the layer. We won't go into the details of all the options, but just to get our map up on the display we can choose to use random colors for each feature or choose fixed colors for the feature (in this case the border) and the area fill. The nice thing is you can set the colors and click the Apply button to see the result without leaving the dialog. This allows you to tweak the colors to get the look you want before clicking OK.

The *Lines* tab allows us to set the line width or use a column in the attribute table of the map for setting line width (you have to set this up in advance of course). Again, you can click Apply to see the effect as you make changes.

The *Symbols* tab allows us to set marker type. Since this is a polygon layer and we aren't displaying the centroid we don't need to set anything here. If we had loaded the `cities` layer this is where we would select the marker style.

The next tab is *Labels* where we can choose to label features using a column name. We can also set the color and background color for the labels, as well as font, label size, and justification. If we want to display labels, we need to choose *Display selected attribute based on 'attrcol'* on the *Required* tab, centroids from the *Selection* tab, and set the name of the column to be displayed on the *Labels* tab. As always, consult the online help for a full description of all the options.

There is one other tab named Optional which contains some rarely used options (at least for me)—you can check out those on your own.

Once we finish making our choices and hit either Apply or OK, we should see the map drawn on the *Map Display* window.

If your map doesn't show up, make sure the layer is selected in the *Layer Manager* and click the `Zoom to selected map layer(s)` tool on the *Map Display*. The `world_borders` map will be drawn to fill the display window. This gives us a simple display of our map. If we want to symbolize the map by population or some other attribute, it turns out you can't do that by adding a vector layer to the map—you have to use a thematic map layer.

Adding a thematic map layer is done in a similar fashion as a regular vector layer. In fact, the button to add a thematic layer is just to the right of the button we just used—if in doubt, hover the mouse. Clicking the `Add various vector map layers` button adds a "thematic 1" object to the manager window. Clicking it brings up the options panel. First we specify `world_borders` as the vector map. You will notice the options for a thematic map are different than for the vector map. The critical thing is we need a numeric field to classify the map with. Fortunately, the population field is precisely that. We can use the drop-down box to set the numeric attribute column to `POP_CNTRY`. We will use graduated colors for our display and set the thematic divisions to "interval". Since we want to be subtle, we will use a custom gradient for the color mapping.

On the *Theme* tab we can set the number of intervals—eight sounds about right for this map. The last tab we will touch is the *Color* tab. Here we want to not draw any outlines so we check the *Only draw fills...* checkbox. The other two colors to set are the beginning and ending colors for the custom gradient. Using 240:240:240 for the beginning and 80:80:80 for the ending will give us a nice grayscale gradient which renders the more populous countries in darker shades.

Once the options are set click Apply or OK and we get the result shown on Figure 18.8.

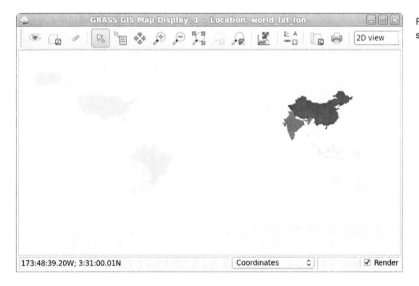

Figure 18.8: GRASS thematic map showing world population

You have probably noticed that the GRASS GUI uses a different paradigm than the other OSGIS applications we surveyed in *Appendix A: Survey of Desktop GIS Software*, on page 299. As such, it takes some getting used to. If you are going to use GRASS, take some time to familiarize yourself with the interface. In fact, let's do some of that right now.

18.4   Getting to Know the GUI

One of the barriers to using GRASS is psychological. Many people dismiss it as being too difficult, arcane, or archaic. That's unfortunate because GRASS has a lot to offer. That's not to say there isn't a learning curve with GRASS. Just like any powerful application, it takes time to learn and master. That said, let's take a look at the map interface in Figure 18.8, on the previous page and examine both the toolbar and the other features.

MDI or SDI?

As you may have guessed by now, GRASS lets you have multiple windows, which essentially makes it a Multiple Document Interface (MDI) application as opposed to a Single Document Interface (SDI) application. Some folks prefer *MDI*, although modern user interface design thought seems to indicate that *SDI* is preferable. Using an *MDI* application on a small display can be quite painful—plan accordingly.

To acquaint us with the tools available on the map interface, take a look at the following list of each tool and a brief description of its function. The name for each is the same tooltip text you see when you hover the mouse over a tool.

`Display map`
> This tool redraws the map, refreshing only those layers that have been modified.

`Render map`
> Redraws the map, refreshing all layers.

`Erase display`
> Erases the map to the chosen background color (default is white).

`Pointer`
> Switches to the pointer. When the pointer is active, its location is displayed in the status bar at the bottom of the map.

`Query`
> This tool is equivalent to the identify tools we saw in uDig and QGIS. It operates on the active layer (map) and returns information about the layer at the location you click. For a raster this will

be the coordinates of the mouse click and the cell value. For a vector layer, it displays the coordinates as well as the attributes for the feature.

Pan

Pans the map when you hold down the mouse and drag it.

Zoom In

This provides an interactive zoom-in tool. Dragging a box zooms in to the outlined area. A single click of the tool zooms in a fixed amount.

Zoom Out

This provides an interactive zoom-out tool that works like that of QGIS. The display will be zoomed so that it is contained in the rectangle created by dragging the mouse. Clicking the map results in a fixed zoom out.

Zoom to selected map layer(s)

This tool zooms to the extent of all selected map layers.

Return to previous zoom

This should be pretty obvious, but it returns you to the previous view. This tool maintains a history of zoom levels, allowing you to click your way back to where you initially started.

Various zoom options

This tool is really a collection of one-shot tools allowing you to zoom to a saved region, the default region, and a computational region. You can also use the tool collection to set the computation region from the display extent and save the display geometry to a new region. Consult the manual for details on these options.

Analyze map

This is another collection of tools that allow you to measure distances on the map, create a profile on a raster and then plot a cross section. This is useful for example in creating elevation cross sections from a Digital Elevation Model (*DEM*). The other tool in this set can be used to create a histogram of a raster that plots the cell values versus the number of cells.

Add map elements
This toolset allows you to add a scalebar, north arrow, legend, and text layer to the map.

Save display to graphic file
Exports the current map view to a graphics file. PPM/PNM, TIFF, JPEG, and BMP are supported.

Print display
Actually, this tool does more than the tooltip implies. It allows you to send the map to a printer or create a PostScript file.

Mode drop-down
The last thing on the map display is a drop-down box that allows you to enable 2D, Digitizing, or 3D modes. When you select a mode, the interface changes to include additional tools or dialogs.

18.5 Digitizing and Editing

To complete our introduction to GRASS, let's look at how to digitize and edit a layer. We'll create a layer and digitize some lakes off a world mosaic raster.

The GRASS Vector Model

Before we start digitizing, we need to go over a few facts about how GRASS stores and works with vector features. The GRASS vector model is all new at version 6.*x*. Rather than go into all the gory the details now, just be aware of the following:

- Each vector feature has a "category" number that serves as an identifier.
- Vector features can have 0..n category numbers.
- A GRASS layer consists of link(s) from a set of geographic objects to attribute table(s).
- Layers don't contain any geographic features but are linked to them.

- For a feature to be part of a layer, one or more of its category values must appear in the attribute table.

You are probably wondering what the story is with this confusing sounding category and attribute table thing. Basically, it's pretty powerful in that it allows you to link multiple attribute tables to a single feature. Why would we want to do that? Well, in a contrived example, suppose for our lakes layer we want to store physical characteristics such as volume and maximum depth, width, and length. We also want to store information about fish species and their percentage of the total fish population. As you can see, these two goals don't really translate into one attribute table. In the first case, it's a one-to-one relationship, and in the second (fish), it's a one-to-many relationship. Besides that, the two types of data just don't belong together.

In GRASS, we can create two attribute tables and link them to our vector layer. Then we can add records to each, using the vector category as an identifier in the attribute tables. If a lake has no fish, it won't have any records in the fish table. If we draw the "fish" layer, that lake won't show up. Although this may not be the best example, you get the idea.

In fact, you can combine totally unrelated features (data-wise) that are topologically related in a single map and break it out into layers using this scheme. The GRASS documentation uses the example of forests and lakes, but any mapping of land parcels that includes water bodies is a good example. The lakes and parcels are topologically related (share common boundaries) but are different types of features. In conventional GIS, we might represent these with two separate layers (for example, shapefiles). In GRASS, they can live in the same map and be distinguished using categories and attributes.

Digitizing and Editing in GRASS

With the advent of the new wxPython GUI, digitizing in GRASS has changed a bit. To get familiar with it, we'll use the Natural Earth raster and digitize some lakes. If you didn't already load the raster into GRASS, see Section 14.1 *Loading and Viewing Data*, on page 238 for information on obtaining it and loading it.

To begin, start GRASS with the wxPython GUI using your `world_lat_lon` location and the mapset that contains the Natural Earth raster. Load the raster and make sure it is displayed in the *Map Display* window. The drop-down box at the right of the *Map Display* toolbar contains a *Digitize* option. Selecting it will add the digitizing tools to the interface and redraw the map. Now we're ready to create the new vector layer.

On the digitizing toolbar there is a drop-down box that allows you to create a new layer or select a layer to edit. Since we don't have any vector layers loaded, the only option is *New vector map*. When you select it, you are given the opportunity to name the new map and create an attribute table with it. Let's name our new layer "lakes" and click OK to create it and add it to the map display. With the empty layer created we are ready to start digitizing.

The first step is to zoom into the area where we want to work. Let's start with Lake Tanganyika, located in the Great Rift Valley of Africa near 6.5 degrees S latitude, 29.5 degrees E longitude. To help locate it, move the mouse and observe the latitude and longitude shown at the bottom left of the map display window.

For digitizing we need to zoom in close enough to see the detail of the feature. Often this mean zooming in quite close and panning as we digitize along the feature.

We are almost ready to digitize, but first we need to add at least one more column to the attribute table associated with our vector map. To do this, highlight the `lakes` layer in the *Layer Manager* and click on the `Show attribute table` tool in the toolbar to bring up the *Attribute Table Manager*. To add an attribute, click on the *Manage tables* tab.

Enter the field name, and then choose the field type from the drop-down list. For the `name` field for our lakes, we'll use a varchar field with a length of 24. Once you have filled in the field name and data type, click the Add button to add it to the attribute table. In Figure 18.9, you can see the completed attribute table with the new name field. If you wanted to add more columns, you can do so using the same process. When done, close the dialog by clicking the Quit button. Now we have a vector map and associated attribute table ready for digitizing.

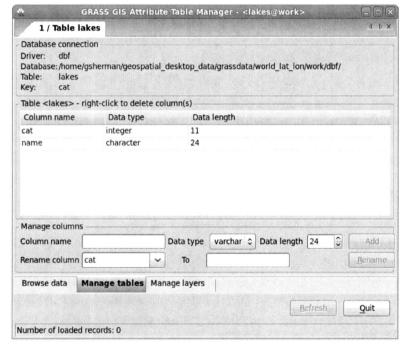

Figure 18.9: Adding a Column to a GRASS Vector Map

To begin editing, select the lakes map and click on the edit tool (looks like a pencil) or right-click on the layer name and choose Start editing from the popup menu. The digitizing toolbar will appear and you are ready to digitize. The tools on the digitizing toolbar allow you to create a full range of feature types as well as edit and delete them. You can also edit the attributes of features and do a nice range of tools for copying, flipping, splitting, and querying features.

For digitizing lakes we only need a few tools:

- Digitize new boundary
- Digitize new centroid
- Digitize new area

We can use the first two to create our lakes, filling in the attribute data for both the boundary and centroid, or we can do it in one operation using the `Digitize new area` tool.

Basically you select to tool, click around the lake following the shoreline and then close the polygon by clicking on the start point and right-clicking. The dialog to enter the attributes pops up and we can enter the name for the lake.

If the boundary turns green when you close it, you're in luck; the start and end points coincide. If not and it's red, we can use the `Move vertex` tool to easily move the last point to close things up. To move it, select the tool and click the last point, and then move the mouse until the cursor is over the first point and click again. If you're lucky (or good), the border will turn red to green, indicating closure.

Now that we have a closed boundary, we can make it a polygon by adding a centroid. Click the `Digitize new centroid` tool, and then click somewhere in the center of the lake polygon. Again the attribute form will pop up, and we can enter Lake Tanganyika for the name. When we submit, the attributes are stored. Now when we click on the lake in the GRASS map display using the `Query` tool, we get the polygon attributes.

In Figure 18.10, on the facing page, you can see the results of our digitizing the lakes. We've rendered them in cyan with a three-pixel dark blue border and added labels to the centroids, all using the *Layer Manager* to set things up.

That completes our simple example of digitizing in GRASS. We didn't get too fancy with things but illustrated a fairly typical session for creating polygons. Of course, lines and points work in a similar fashion. If you need to refine your work, the other tools on

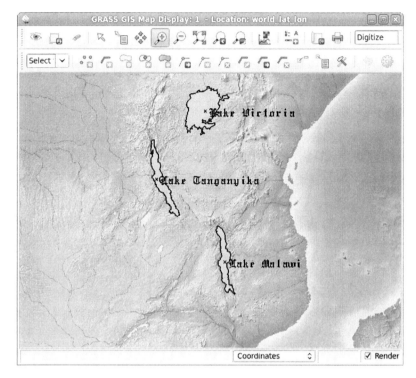

Figure 18.10: Results of digitizing lakes in GRASS

the digitizing toolbar allow us to add and delete a vertex, split a line, and move or delete an entire feature.

Also on the digitizing toolbar, you'll find tools to copy categories for a feature. Why would you want to copy categories from one feature to another? Well, imagine we have two features that belong to the same overall "thing." An example is a parcel of land divided by a river. When you digitize, you will have two polygons, but they both are the same parcel of land—they have the same owner, tax ID, or what have you. We can digitize the first polygon and set its attributes. Then when we digitize the second, rather than assigning a new category number and duplicating all the attributes, we just copy them from the first parcel using the Copy categories tool. So, this tool is good when we are working with multipoint, multilinestring, or multipolygon features.

This section is about digitizing and editing with GRASS, and you're thinking "all we did was talk about digitizing." Well, as with most of the editing-enabled applications, the operations are pretty much the same. How do we edit the features in GRASS? Just start editing from the *Layer Manager* and have at it. We can also manipulate the attribute data using the db.* suite of commands.

19

Appendix D: Quantum GIS Basics

Quantum GIS has a lot of functionality and many areas to explore. Here you will find an introduction to the basics of using QGIS, including map navigation and other essential features.

At first glance, QGIS has a feature set similar to other GIS viewers/editors. QGIS can view vector and raster data, including data stored in a PostGIS-enabled PostgreSQL database. QGIS also supports WMS and WFS layers.

We used QGIS to view and render data in Chapter 5.4, *Advanced Viewing and Rendering*, on page 48, so we won't repeat that here. Instead, we'll take a look at some of the features you may not have seen yet.

19.1 *Vector Properties and Symbology Options*

As you know, each layer you add in QGIS has properties associated with it. You can access the vector properties dialog box by double-clicking the layer name in the legend or by right-clicking its name and choosing Properties from the pop-up menu. In this section we'll take a brief tour through some of the options on the vector properties dialog.

The vector properties dialog has the following tabs:

- Style
- Labels
- Fields
- General
- Metadata
- Actions
- Joins
- Diagrams

There is a lot to the vector properties dialog—let's take a look at some of the key features.

Style

The Style tab is where you set the symbology for a vector layer. QGIS has two options for symbology—new and old. By default QGIS uses the new symbology and you can switch to the old by clicking on the *Old symbology* button.

One of the main features of the new symbology is the level of control you have over the look of your symbols and the ability to create and save custom styles. At version 1.7.x the only reason I find to use the old symbology is when I need a unique value renderer, such as we used in Chapter 5.4, *Unique Values*, on page 56.

Both symbology methods allow you to set the transparency of your layer and to save the styling to a .qml file for future use.

Let's look at a few details.

New Symbology

The new symbology Style tab contains the following:

Legend Types
   The drop-down box contains a list of legend types. These are actually ways to symbolize your data. In the new symbology you can choose from Single Symbol, Categorized, Graduated, and Rule-based renders.

Symbol levels

Sometimes you want to control the order in which symbols are drawn when using one of the renders. For example, if we rendered earthquakes using a color and/or larger symbol size for major quakes, we would want these to be drawn last so they aren't obscured by smaller quakes. The *Symbol levels* button opens a dialog that allows you to control the draw order.

Unit

This drop-down specifies the units for drawing features. You can choose from millimeters or map units.

Transparency

The transparency slider allows you to make the layer transparent. This allows you to "see through" the layer to reveal layers lower in the map stack.

Color

The color *Change* button allows you to quickly change the color for the features in the layer.

Change

The *Change...* button allows you to completely customize the look of the symbol, including:

- Fill type
- Fill style (for example solid, diagonal, etc.)
- Marker type
- Line type
- Color
- Border color
- Border style (for example solid, dashed, dotted)
- Border width - the width of the border in the units you chose on the *Style* tab. The options you'll see depends on the geometry type of the layer (point, line, or polygon).
- Offset for the symbol in the X and Y direction. This will shift the drawing of the symbol from its actual location.

Advanced

The *Advanced* drop-down allows you to use data defined rotation and size scaling using fields in your attribute table.

Save as style

This button saves the style so that it will be available whenever you open the *Style* tab in new symbology. This is different than saving the style to a .qml file using the *Save Style...* button at the bottom of the vector properties dialog.

Style Manager

The Style Manager allows you to manage the styles for markers, lines, fills, and color ramps. You can export and import styles in XML format. You can import new styles created by others as long as they are saved in the proper XML.

Old symbology

This button switches to the old symbology dialog.

Old Symbology

The options you'll see (marker type, line type, fill type, etc.) on the Old Symbology dialog also depend on the geometry type.

Legend Type

When you choose to use the old symbology, you have access to Single Symbol, Graduated Symbol, Continuous Color, and Unique Value renderers. For examples of using these to render your data, refer to Chapter 5.4, *Advanced Viewing and Rendering*, on page 48.

Transparency

The transparency slider allows you to make the layer transparent. This allows you to "see through" the layer to reveal layers lower in the map stack.

Label

The Label field allows you to specify the name that appears next to the symbol in the legend. So for the world_borders layer, we might name it "World Borders."

Point Symbol

This drop-down contains the available symbols for a point. When you select one from the list, it will be used to render the points on the layer.

Point Size

This allows you to change the size of the symbol. The size is specified in points just as font sizes are in your word processor.

Outline Style

For line and polygon layers, you can specify the style of the line used to draw the feature or its outline. This also works with the basic point symbols (box, triangle, circle, and diamond). QGIS provides a number of style choices, including solid, dashed, dotted, and various combinations of dot-dash.

Outline Color

You can specify the color of the outline for the basic point symbols as well as for line and polygon layers. Just click the colored box to the right of the option to open the color selector.

Outline Width

This option lets you specify the width of the line in pixels (used for line and polygon layers).

Fill Color

For a polygon layer, you can specify the color used to fill the polygons or basic point symbols. Again, just click the colored box to the right of the option to open the color selector.

Fill Patterns

For polygon and basic point symbol fills, the default is a solid. QGIS has a range of other fill patterns you can choose from (including hollow). Just click the pattern you want to use to select it.

Labels

On the *Labels* tab you'll find what we refer to as "old" or original labeling engine for QGIS. The new labeling engine can be accessed from the Labeling toolbar (look for the *ABC* tool).

In the simplest case you go to the *Labels* tab and check *Display labels*, then choose the field in the attribute table that you want to use. This is enough to get labels on the map canvas. From there you can go on to adjust the font settings and the placement options. You can also control when labels will be visible using scale dependent rendering. The offset options let you control how far from the label point the text will appear.

On the *Advanced* tab of the *Labels* tab you'll find everything you need to control labeling using fields in the layer attribute table. Of course you need to have prepared your data in advance to support these features.

Fields

The ability to delete a field may depend on the type of vector layer (shapefile, PostGIS, Spatialite, etc.). GDAL/OGR version 1.9 adds support for deleting fields from shapefiles.

The *Fields* tab lists the fields in the attribute table of your layer, but it is more than just a list. When you toggle edit mode you can add and delete columns, as well as use the field calculator to populate or update field values. The field calculator allows you to create a new field and set the values for it, or update an existing field. A simple example would be a `cities` layer containing an `ELEVATION` field with values in meters. If we also needed a column with elevation in feet, we could use the field calculator to create the column and populate it using a formula of `ELEVATION * 3.28`.

You can also set the edit widget type for each field. Typically this is just a regular text edit box. For some fields it may be helpful to use other options, such as unique values (a drop-down box), a checkbox (for booleans), file name, or calendar. In our `cities` layer we could set the `CAPITAL` to use a *Unique values* editor to set the value to 1 for yes and 0 for no. In order for these settings to be remembered you must save the changes using the *Save Style...* button. This will create a .qml file (for example cities.qml) that stores not only your field customizations but all the styling, including classification and color(s).

General Tab

The *General* tab contains options and settings that are of all things—"general." First off we see that we can define a display name for the layer. You can do this by right-clicking the layer name and choosing Rename. That's the quick way to do it, but you can also set it here if you like.

The *Display name* controls the field that is used to summarize Identify results and display Maptips.

The Edit UI box allows you to specify a Qt Designer form that can be used when editing and creating new features. This not only gives you a custom look and feel, but you can add validation code by specifing a Python Init function. Using a custom user interface can make your data capture routines more robust as well as more pleasing to the eye.

For an example of using a custom UI, see the Resources at http://geospatialdesktop.com/resources

Scale-Dependent Rendering   QGIS supports *scale-dependent rendering*, allowing you to control when a layer is visible. There are a couple of primary reasons why would we would want to do that:

- The layer is meaningless at small scales (remember, a small scale covers a large area) because you wouldn't be able to see the information. An example is displaying streets at the scale of our world_borders layer. You wouldn't gain much information from such a display.
- You have two versions of the data, one for small scales and the other for large. An example of this is a coastline. At a small scale (zoomed way out), we don't need to display a lot of detail. In fact, such detail is wasted and only slows the rendering process. You can cram only so much information into a single pixel. As you zoom in, you want the low resolution layer to switch off and

another higher resolution (more detail) layer to switch on.

If you choose to use scale-dependent rendering, you have to set the minimum and maximum scales for the layer. This controls the visibility of the layer. The scale for each setting is specified as a ratio of 1:[some number]. How do you determine what that number should be? Fortunately, QGIS displays the current scale in map units on the status bar. You can use this information to determine what scales you want for the minimum and maximum settings. Use the zoom tools to get the view you want at each level and then use the displayed scale to set up the scale-dependent rendering.

What if we want the layer to always be visible when zoomed out (small scales) but turn off as we zoom in? Set the scale you want it to turn off at in the *Minimum* field, and set the *Maximum* field to a very large number. If you want the inverse to be true, set the *Minimum* field to zero and the *Maximum* field to the scale at which you want the layer to turn off.

Note that the terms *minimum* and *maximum* as used in the dialog box seem to be opposite of our definition of small vs. large scale. As long as you are aware of this, you'll be able to set things up to meet your needs.

Spatial Indexes   The next area of interest on the *General* tab is the *Create Spatial Index* button. From here you can create a spatial index, assuming the layer type supports such an option. A spatial index speeds up drawing, selecting, and identifying features. For example, when zooming in, QGIS uses the spatial index to select only the features in the view window for drawing. This is much quicker than marching through each feature in the layer and testing it to see whether it should be drawn. The same holds true for selecting or identifying (which really involves a select of sorts) features. The spatial index helps quickly locate the feature(s). When you click the button to create a spatial index for a shapefile, a new file with a `qix` extension is created. For our `world_borders` layer, the spatial index file is `world_borders.qix`. Always make sure to cre-

ate a spatial index for layers that support it. It makes things much
snappier.

Coordinate Reference System   The *coordinate reference system* defines
the projection and coordinate system of the layer. This is what
makes it possible to draw data in real-world coordinates and have
differing layer's "line up." The projection information for the layer
is displayed in EPSG or PROJ.4 format (see Chapter 11, *Projections
and Coordinate Systems*, on page 159 for details). Using the *Specify
CRS* button, you can set the projection for the layer.

Usually QGIS sets the parameters you see here based on the projection information associated with the layer. If there is no projection
information, then QGIS makes an assumption based on your preference setting for projections. You can set QGIS to do the following
when a layer is loaded that has no projection information:

- Prompt for CRS
- Use the project CRS as set in the Project Properties
- Use the default CRS specified in your user preferences

If you want to take a look at these options now, open the *Preferences*
dialog box, and click the *CRS* tab.

Subset   The last piece of the *General* tab concerns itself with the
subset of the data. This feature lets you create a subset of your
layer on the fly and display only the data matching the criteria you
specify. For an example of how to use this feature, take a look at
Section 9.4, *Using PostGIS and Quantum GIS*, on page 124.

Metadata Tab

The *Metadata* tab provides detailed information about the layer, including its location, type, number of features, and projection. In
Figure 19.1, on the following page, we can see the *Metadata* tab contents for our world_borders layer.

Figure   19.1:   Metadata   for   the
world_borders layer

The General information provided shows us not only information about the layers physical location but also the type—in this case an ESRI shapefile. We can also see that our layer contains polygon features and has a number of editing capabilities. Well, these aren't really capabilities of the layer but show you what QGIS can do with the layer in terms of editing. With the `world_borders` layer, we see that the capabilities include add and delete features, create a spatial index, and change the values for attributes in the layer. This is useful information, especially in the case where you just downloaded a new layer and it's still a mystery to you.

The metadata also includes information about the extent of the layer. It shows the extent in both the coordinate system (CRS) of the layer and the project. You might notice that they are the same in our example. This is because we haven't changed the coordinate system of the map canvas. We are displaying the `world_borders` features in the same coordinate system in which they are stored.

The layer and project coordinate systems are displayed in the next section of the metadata, using PROJ.4 format. In this case, it's `+proj=longlat +ellps=WGS84 +datum=WGS84 +towgs84=0,0,0,0,0,0,0 +no_defs`. QGIS isn't very friendly in describing the projection in terms that we can easily understand. If you are familiar with GIS and coordinate systems, you can probably guess that the coordinates are latitude and longitude in the WGS84 datum. Refer to Chapter 11, *Projections and Coordinate Systems*, on page 159 for hints on how to decipher what projection you're working with.

Actions

Attribute actions are handy things that allow you to call an external program and pass it values from the layer's attribute table. The *Actions* tab is where you create and manage actions for a layer. You can get an in-depth look at them in Section 5.5, *Using Attribute Actions*, on page 68.

368 CHAPTER 19. APPENDIX D: QUANTUM GIS BASICS

Joins

The *Joins* tab allows you to join a spatial layer with another layer or attribute table. To successfully create a join you must have a common field in each member of the join. Joins are created by loading the layers or layer and data table in QGIS using the Add Vector Layer... menu and then opening the vector properties dialog. Clicking the plus tool on the *Joins* tab give you a dialog where you can select the join table and field.

Once joined, the attributes show up when identifying features on the spatial layer or when viewing the attribute table.

Diagrams

The Diagrams tab allows you to place pie charts or text diagrams on the map canvas for each feature. This can be useful when you want to graphically convey additional information about a feature. See the QGIS reference manual for details on using diagrams.

19.2 *Project Properties*

By choosing Settings→Properties... from the menu you can customize a number of properties for a QGIS project including:

- Project title
- Selection color
- Background color
- Save paths (absolute or relative)
- Layer units
- Precision

The project title shows up in the title bar of QGIS. If you haven't set a title for a saved project, the file name without the .qgs extension will be displayed in the title bar.

The default selection color is yellow and the default background color is white. Both can be customized as you see fit.

When you save a QGIS project, the full path to each of the data layers is stored in the project file. This is fine, but makes your project not very portable. By setting *Save paths* to relative, all paths are stored relative to the location of the project file. This makes it easier for you to distribute a project file with data.

You can set the layer units for a project to meters, feet, decimal degrees, or degrees, minutes, seconds (DMS). This setting is only valid when on the fly transformation is not set.

The *Precision* setting controls how many decimal places are shown in the QGIS status bar. If you want to customize it, click the *Manual* option and set the number of decimal places.

19.3   *Map Navigation and Bookmarks*

QGIS supports the typical map navigation toolset with zoom and pan. We can also create spatial bookmarks that allow us to store a view extent and return to it later.

Zooming

Load a layer, click the `Zoom In` tool (looks like a magnifying glass with a plus sign), and then drag a rectangle on the map canvas to zoom in. If you can't find the zoom tool, hover the mouse over the tools in the toolbar, and check out the tooltip help that pops up for each one. You can also zoom in by a fixed amount by just clicking the map canvas. Zooming out works the same way but with one nuance. When you drag a rectangle to zoom out, the current view will be scaled to fit in the rectangle.

You can also use the mouse wheel to zoom in and out on the map. First, make sure the mouse cursor is over the map canvas. To zoom in, roll the mouse away from you (think of it as flying toward the map). To zoom out, roll it toward you (flying away from the map). The focal point of the zoom is the location of the mouse cursor. After a mouse zoom, the map will be centered at the cursor location. You can change this behavior on the *Map Tools* tab of the `Settings→Options...` dialog box. If you like, you can set it to just zoom in and not re-center the map. You can also set the zoom factor used with each "roll" of the wheel. Using the mouse wheel to zoom can be very useful when digitizing data.

We need to mention a several other zoom tools here:

Zoom Full

This tool zooms to the full extent of the layers on the map canvas. This means you will see all the features in all the layers on the map. This is a quick way to reset the view after you have zoomed in and panned around.

Zoom to Layer

This zooms to the extent of the currently selected a layer. The selected layer is the one highlighted in the legend. You can invoke this by using the tool button or by right-clicking the legend and choosing Zoom to layer extent from the pop-up menu.

Zoom to Selection

In QGIS (as in most GIS applications), you can select features in a layer either interactively on the map using a selection tool or from the attribute table. If you have a selection set, you can zoom to just the area it covers using this tool.

Zoom Last and Zoom Next

These two tools allow to move backward and forward through the zoom history. This can be a quick way to change the view rather than having to manually perform the zoom operation.

Panning

Panning is done using the Pan tool on the toolbar. It looks like a hand and is located in the same toolbar as the zoom tools. Again, use the mouse hover and tooltips to help determine the function of any tool on the toolbars. To pan, select the tool, and drag it on the map canvas to move the view.

You can also pan using the mouse and without clicking the map canvas. This works only if the map canvas has the focus, meaning you haven't just worked with some other component of the interface, such as the legend. To pan, hold down the spacebar, and move the mouse. This method of panning is very useful when digitizing data, since switching to the Pan tool may have undesired consequences.

Spatial Bookmarks

Bookmarks are everywhere—in your web browser, text editors, and even books. QGIS supports spatial bookmarks, a way to save a location and return to it later. Spatial bookmarks are stored globally, meaning they are available in QGIS regardless if you have the same layers loaded as when the bookmark was created.

To create a bookmark for an area you are viewing, simply click the New Bookmark tool (its the half-globe with star above it), or choose New Bookmark from the View menu. This opens a simple one-line dialog box that allows you to enter a bookmark name. Click OK, and the bookmark is created. Make sure the name you create is descriptive. You might also want to append the coordinate system to the name as a reminder. The bookmark tool doesn't store the projection (it just stores coordinates), so it's up to you to remember or annotate it if it's important.

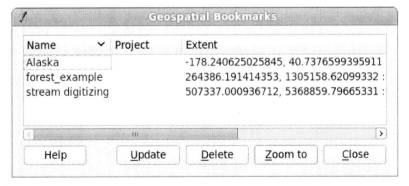

Figure 19.2: QGIS Geospatial Bookmarks dialog box

We can manage our bookmarks using the *Geospatial Bookmarks* dialog box shown in Figure 19.2. Note we have three bookmarks stored. The name and coordinates (Extent) are visible. Since bookmarks are global, QGIS doesn't currently store anything in the Project column. To zoom to a bookmark, select it from the list, and click the *Zoom To* button. You can also zoom to a bookmark by double-clicking it in the list.

You can update the extents of a bookmark by zooming to the new extent, selecting the bookmark name, and clicking the *Update* button. You can also delete a bookmark by selecting it and clicking the *Delete* button. There is no way to rename a bookmark.

19.4   *Plugins*

As you may have gathered already, QGIS supports the use of plug-ins to add new capabilities and tools. Basically, a plugin is a load-able module that can be added and removed from QGIS. QGIS sup-ports plugins written in both C++ and Python.

QGIS has a lot of plugins that are installed by default. Let's get a brief description of some of the commonly used plugins distributed with QGIS.

Add Delimited Text Layer

Loads and displays data from a delimited text file. The text file must contain X and Y coordinates for each feature. Only point data is supported by the plugin. You can find an example of using this plugin in Section 10.2, *Importing Data with QGIS*, on page 141.

Coordinate Capture

Captures mouse coordinates in a CRS different than that of the data. You can select the CRS you want to capture coordinates and copy the results to the clipboard.

Copyright Label

Displays copyright information on the map canvas. You can cus-tomize the text, style, and placement of the label.

GPS Tools

Tools to load and import GPS data, as well as download to your

GPS unit. For details on using this plugin, see Section 10.4, *Using GPS Data with QGIS*, on page 149.

GRASS

A full suite of GRASS tools for loading vector and raster maps, digitizing features, and using modules to import, export, and process data. We cover this plugin in a fair bit of detail in Chapter 14, *Getting the Most Out of QGIS and GRASS Integration*, on page 237.

GdalTools

This plugin integrates the GDAL suite of raster tools into QGIS. These are available from the *Raster* menu once the plugin is enabled.

Georeferencer

Georeference rasters by interactively setting control points to create a world file. See Section 10.5, *Georeferencing with QGIS*, on page 155.

North Arrow

Displays a customizable north arrow on the map canvas. You can adjust the placement of the arrow, as well as the angle. Or you can just let QGIS determine the angle direction automatically.

Offline Editing

Allows offline editing and synchronizing with a database.

OpenStreetMap

Provides a viewer and editor for OpenStreetMap data.

Plugin Installer

This plugin provides the interface needed to download and install Python plugins from a number of repositories.

SPIT

The Shapefile to PostGIS Import Tool (SPIT) allows you to import shapefiles into PostGIS. The use of SPIT is covered in Section 9.3, *Using QGIS to Load Data*, on page 121.

374 CHAPTER 19. APPENDIX D: QUANTUM GIS BASICS

Scale Bar

Displays a scale bar on the map canvas. You can customize the placement, style, color, and size of the scale bar.

WFS

Plugin for consuming WFS services on the Internet and displaying the data in QGIS.

eVis

An event visualization tool that allows you to view images associated with vector features.

fTools

A suite of tools for vector data analysis and management. These tools are available from the *Vector* menu when the plugin is enabled.

QGIS plugins are managed by the *Plugin Manager*. You access the *Plugin Manager* from the Plugins menu. In Figure 19.3, on the next page, you can see the *Plugin Manager* and some of the plugins distributed with QGIS.

The *Plugin Manager* is easy to use. Basically, you just check the plugins you want loaded and click the OK button. If you want to unload a plugin, uncheck it before you click OK. That's pretty much it when it comes to managing plugins. You probably noticed the *Filter* text box at the top of the manager dialog box. This allows you to narrow the list of plugins by displaying only those that contain the filter text. This becomes handy when you have dozens of plugins installed.

When you load a plugin, the plugin's menu and toolbar icons are added to the QGIS GUI. Generally the plugin icons are added to the Plugins toolbar and menus are added to the main Plugin menu item. Not all plugins follow this rule; some may add custom menus and toolbars. An example is the GRASS plugin. It has ten tools and therefore its own toolbar. Another example is the fTools plugin; it adds a Vector menu item to the main menu bar.

Figure 19.3: QGIS Plugin Manager

Creating a Plugin

Creating a plugin in QGIS can be a bit of a technical affair. Creating a new plugin requires the use of C++ or Python and Qt.

Most new plugins are being written in Python because of the relative ease in coding compared to C++. You can install the Plugin Builder plugin to get you started with a working template. For an example of a QGIS plugin written in Python, see the handy Zoom To Point plugin we created in Section 15.2, *A PyQGIS Plugin*, on page 273.

If you really want to write a C++ plugin, check out the resources available on the QGIS website, blog, and wiki.

With the advent of the Python bindings for QGIS, a good number of Python plugins have emerged. At the time of this writing, there

were 180 Python plugins available in the Plugin Installer. That's
a lot of tools for you to checkout and experiment with. For more
information, check out the QGIS Python Plugin repository.[1]

[1] http://pyqgis.org

20

Index

CPSIA information can be obtained at www.ICGtesting.com
Printed in the USA
BVOW062200210513

321312BV00004B/81/P

9 780986 805219